ONLY A THEORY

ALSO BY KENNETH R. MILLER

Finding Darwin's God

KENNETH R. MILLER

ONLY A THEORY

Evolution
and the Battle for
America's Soul

VIKING

VIKING
Published by the Penguin Group
Penguin Group (USA) Inc., 375 Hudson Street,
New York, New York 10014, U.S.A.
Penguin Group (Canada), 90 Eglinton Avenue East, Suite 700, Toronto,
Ontario, Canada M4P 2Y3 (a division of Pearson Penguin Canada Inc.)
Penguin Books Ltd, 80 Strand, London WC2R 0RL, England
Penguin Ireland, 25 St. Stephen's Green, Dublin 2, Ireland
(a division of Penguin Books Ltd)
Penguin Books Australia Ltd, 250 Camberwell Road, Camberwell,
Victoria 3124, Australia (a division of Pearson Australia Group Pty Ltd)
Penguin Books India Pvt Ltd, 11 Community Centre,
Panchsheel Park, New Delhi–110 017, India
Penguin Group (NZ), 67 Apollo Drive, Rosedale, North Shore 0632,
New Zealand (a division of Pearson New Zealand Ltd)
Penguin Books (South Africa) (Pty) Ltd, 24 Sturdee Avenue,
Rosebank, Johannesburg 2196, South Africa

Penguin Books Ltd, Registered Offices: 80 Strand, London WC2R 0RL, England

First published in 2008 by Viking Penguin, a member of Penguin Group (USA) Inc.

1 3 5 7 9 10 8 6 4 2

LIBRARY OF CONGRESS CATALOGING-IN-PUBLICATION DATA
Miller, Kenneth R. (Kenneth Raymond), 1948– .
Only a theory : evolution and the battle for America's soul / Kenneth Miller.
p. cm.
Includes bibliographical references and index.
ISBN 978-0-670-01883-3
1. Evolution (Biology)—Study and teaching—United States. 2. Intelligent design (Teleology)—Study and teaching—United States. 3. Science—United States—Psychological aspects. I. Title.
QH362.M55 2008
576.8071'073—dc22 2008013918

Printed in the United States of America

To my daughters, Lauren and Tracy,
lovers of Nature, Life, and Learning

Contents

Preface

The words belonged to Disraeli, but their delivery was pure William F. Buckley, Jr., dean of American conservatism and iconoclastic critic of all things left and liberal. At the end of a rousing televised debate on evolution, Buckley invoked the words of the British statesman to sum up his own views on Darwin: " 'What is the question now placed before society? The question is this: Is man an ape or an angel? I, my Lord, am on the side of the angels.' "

Buckley's erudite summation, in that 1997 debate on his program, *Firing Line,* was no mere turn of phrase. Buckley viewed Disraeli's decision to stand with the angels as far more than a matter of mere preference, assuring his opponents "that beneath their surface glitter and hidden within their metaphor, these words contain a truth that shall someday break the pieces of the new philosophy, which [Thomas] Huxley spent his life so devotedly to establish." That "new philosophy," which Buckley hoped to smash, was evolution.

I was one of several who stood in opposition to Buckley's views that evening, and I marveled that an intellect as strong as his could possibly have rejected the cornerstone of modern biology, but reject it he did. In time, I came to see that his opposition, and that of many other Americans, to the theory of evolution was symptomatic of a deep unease within our society over the nature of modern science. For them, evolution is far more than a mistaken scientific theory. It is the cutting edge of a dangerous and destructive materialism that threatens the heart and soul of our civilization and culture, and it must be opposed at all costs.

In many ways, I think that William F. Buckley misunderstood evolution, and he certainly misunderstood the nature of science. There is indeed a soul at risk in America's "evolution wars," but it is not the cultural one that Buckley sought to save. Rather, it is America's scientific soul, its deep and long-standing embrace of discovery, exploration, and innovation, that is truly at risk. That is why the stakes in this struggle are far greater than the wording of curriculum standards or the nature of textbook passages on the Cambrian fauna. The choice we face as a nation is nothing less than whether we will continue to be the world's scientific leader or quietly watch as the torch of discovery is lifted from our hands. I wrote this book to describe the nature of that choice and the realities of the struggle, and above all in the hope of reaching those who, like the late William F. Buckley, have yet to appreciate that the great gems of Western culture and thought are inextricably linked to scientific reason and rationality.

Many individuals contributed to the ideas presented in *Only a Theory,* and I am grateful for their help, their support, and most especially their friendship. These include Thomas Schneider, who was kind enough to discuss information theory with me and to review part of my manuscript; David DeRosier, a patient and supportive scientific friend whose devotion to the art and craft of biology has always inspired me; and Francis Collins, whose thoughts on the relationship between science and faith have helped open the eyes of many who assume that these two ways of knowing must always be in conflict. I am grateful for the support of many wonderful people who share my views on the importance of this struggle, including

Lawrence Krauss, Fr. George Coyne, Steve Case, Patricia Princehouse, Eugenie Scott, Nick Matzke, and Keith Miller.

I was fortunate to have played a small role in the landmark *Kitzmiller v. Dover* trial in 2005 and will remain ever grateful for the opportunity. In so doing, I was able to work with people of exceptional courage and conviction, such as Dover Area High School teachers Bertha Spahr, Jennifer Miller, and Rob Eschbach, as well as Tammy Kitzmiller and her fellow plaintiffs. I remain in awe of the legal talents of attorneys Vic Walczak, Eric Rothschild, and Steve Harvey and am honored to have joined fellow witnesses Barbara Forrest, Jack Haught, Kevin Padian, and Brian Alters in this extraordinary trial. For better or for worse, we will be linked forever in a case that struck an important blow for science education and religious freedom.

I thank my colleagues at Brown University for their continuing support and many kindnesses as I worked on this book, as well as my students and advisees for their forbearance in the face of my efforts to gather material for this manuscript. Finally, and most important, I thank my wife, Jody, for her love and understanding, and my wonderful daughters, Lauren and Tracy, to whom this book is dedicated.

ONE

Only a Theory

IN A COURTROOM even a whisper can catch your attention, especially one that comes right at you with a smile and a wink.

"Only a theory," she said, shaking her head just enough to get my attention as I walked past her. "It's only a theory—and we're gonna win." Her smile was genuine, and its certainty was unmistakable.

She didn't win—at least not that day and not in that court—but the quiet confidence of the remark has stayed with me ever since. It has provoked me to doubt, wonder, and even fear—and it's my inspiration for writing this book. It came at the conclusion of my first trial, the first time I'd ever sat in a witness stand and given testimony, the first time I'd ever been cross-examined, the first time I'd ever had to meet reporters on the courthouse steps. But it wouldn't be the last—and that's part of this story, too.

When I walked into a federal courtroom in Atlanta in the fall of 2004, I could easily have concluded that a book was on trial. An attorney stood next to a four-foot-high enlargement of the table of contents of a biology textbook. Nearby a collage of more than fifty pages from the evolution section of the same book had been pasted on cardboard and placed on a mounting stand. It seemed to form a wall of evidence that might be used, one could suppose, to convict the book or its authors of some awful, seditious offense against the state or against the good schoolchildren of Georgia, for whom the book had been written.

These were first impressions, to be sure, but they were impressions that mattered, especially to me, the coauthor of that book. It was almost as though the project on which Joseph Levine and I had labored for so many years had been cut apart, and now its entrails were glued to that board like the organs of a laboratory animal pinned against the soft wax of a dissecting tray. When I climbed into the witness stand, I wondered if the attorneys regarded me that way, too. Were they looking at me as I might look at a laboratory mouse? Trying to find a quick and easy way to get inside and take what they needed?

I would find out soon enough. But the remarkable thing about that trial—apart from the packed courtroom, the media attention, the calls from reporters—was the size and scale of what was being litigated. All of the effort, paperwork, and argument was focused on a paper sticker, about six inches square, containing just three sentences:

This textbook contains material on evolution. Evolution is a theory, not a fact, regarding the origin of living things. This material should be approached with an open mind, studied carefully, and critically considered.

Why such a fuss? The issue in this trial was whether the actions of the Cobb County Board of Education, in affixing this sticker to the inside of thousands of public school science textbooks, amounted to a violation of the First Amendment of the United States Constitution. Constitutional questions are always matters of great interest, and one that applied to the public schools, where most American children are educated, naturally drew plenty of notice and passion.

In the end the court found that the stickers were indeed such a violation and ordered them removed,[1] but it didn't take a psychologist to sense that there was something more at work here, something far deeper than the establishment clause or the narrow scientific meaning of a word like "theory." The stickers were actually the result of a school board's effort to fashion a compromise between thousands of its constituents and the science education standards that their schools were required to meet. Those standards had pushed evolution, the central organizing principle of the biological sciences, into the textbooks, classrooms, and even the homes of the families of Cobb County, and thousands of them had pushed back.

They resisted with sermons from pulpits around the county, with protests at school board meetings, and with a massive petition drive demanding that the board either remove evolution from the classroom or teach it alongside an alternative more friendly to their own views of God's creation. You might say that the sticker was the board's half-baked effort to split the difference between God and science—and that it ultimately satisfied neither.

I had a great deal to say about the wording of that sticker during my testimony, but what most impressed me at the trial was the passion of those who defended it. That passion inspired a civic movement—not just in Georgia, but all across the country—and led a majority of Americans to state that they rejected an idea at the very heart of biology, the theory of evolution. To them, what was at issue was a question of the heart and soul. They were prepared to fight for that issue, and fight they would.

A BATTLE IS JOINED

As I left the courtroom in Atlanta, something made me recall the last time I had seen such certainty in the face of defeat. It was the summer of 2000, and I had just driven for hours across the scorching plains to speak to several hundred people in the basement of the First Lutheran Church in Manhattan, Kansas. I was on a crusade.

A year earlier the Kansas Board of Education had fired an opening salvo by excising all mention of evolution from the state's new public school science curriculum. Science teachers in the state were stunned and embarrassed, and they resolved to do something about it. Before long a loose coalition of teachers, science professionals, and other citizens had formed to reverse the board's actions in the most direct way possible—by voting its members out. That was precisely the reason for my July "holiday" in the Midway State. Most members of the board would soon be running in primary elections, and this particular week was, ironically, the anniversary of the Scopes trial. What better time to run a series of teach-ins on the issue of evolution exactly seventy-five years after Clarence Darrow and William Jennings Bryan had done battle over the same issue in a steamy courtroom in Dayton, Tennessee?

The summer of 2000 was hardly the first time Americans had come to Kansas to do battle over an idea. In the 1850s, concerned lest they find themselves in a situation where free states might have a majority of votes in the United States Senate, slaveholding states fashioned a new political bargain that undid decades of careful compromise. The Kansas-Nebraska Act of 1854 threatened to upset a delicate balance in the Senate between free and slaveholding states by allowing the issue of slavery in both states to be decided by popular sovereignty. The act led to a sudden influx of abolitionist and proslavery forces into the territories, and skirmishes over the issue quickly came to dominate the affairs of both emerging states. Proslavery and antislavery forces fought a series of bloody skirmishes across Kansas, and at one point the battle became so intense that the antislavery town of Lawrence was burned to the ground.

Even today history books speak of "bloody Kansas," a burned-over district in which abolitionist jayhawkers clashed with proslavery bushwhackers. Murders and atrocities on both sides would divide the state and eventually the nation into two camps. What took place there was, in hindsight, a savage rehearsal for the Civil War.

But this wasn't civil war, I reminded myself, and in truth, the evening in Manhattan, Kansas, went well.

Just as the timing was perfect, the issues, at least on one level, seemed perfectly clear. Simply stated, did Darwin get it right? Throughout the earth's history, had species leaped suddenly into existence, or had they descended, as the old man wrote, with modification, from an endless chain of ancestors? Was the earth truly old enough to have allowed natural selection sufficient time to produce the remarkable adaptations that fit species so beautifully to their environments? Could the mechanisms of molecular genetics actually have produced the biochemical and physical novelties that evolution demands?

The answers to such direct questions are not hard to come by. In almost every respect Darwin *did* get it right, and that was the message I intended to bring to my audience. The very ground upon which we stand is eloquent testament to the age of this planet. The expanding richness and diversity of the fossil record documents one case after another of descent with modification. Our expanding knowledge of molecular genetics completes exactly the mechanistic framework for which Darwin might have hoped.

Explaining all of this promised to be great fun, and so it was. Science at its best is neither an acquired taste nor an exercise in an exclusive form of mental gymnastics. Properly explained, science is nothing more than organized common sense, and that's exactly what I did my best to bring along that evening. As I will explain later in this book, the specific objections raised against evolution are easily answered, and I had no trouble answering them in Manhattan. For many of the people in my audience that night, that was enough. They had come to see for themselves if evolution was the "dying theory" that its opponents had claimed. When case after case of evidence to the contrary was laid before them, they came to the same conclusion as did the scientific community: that evolution has never been on stronger scientific ground than it is today.

But for a handful in the audience, nothing I could ever say about evolution would be sufficient, absolutely nothing. To be sure, these folks might bring up certain scientific issues in the battle against evolution, but the issues themselves were not their real problem. There was something in the science itself that bothered them, giving rise to an undercurrent of unease and fear that no fossil, no DNA sequence, no experiment in lab or field could ever address. And they were certain, certain beyond belief, not just that evolution was not the answer, but that evolution *could not be the answer.* These were exactly the people whose quiet self-assurance reminded me so much of my new "friend" in Atlanta.

The opponents of evolution lost the 2000 elections in Kansas, and a new, proscience board took office the following year. The victory was gratifying but temporary. Just as in bloody Kansas of the 1850s, it wasn't long before territory claimed by one side was reclaimed by the other. A 6–4 antievolution majority took back the Kansas board in the 2004 elections, and the following year they once again rewrote the science standards to their liking. This time, however, they didn't remove evolution from the curriculum, but introduced evidence "against" it and redefined science in a way that made practicing scientists and educators shudder. Once again proevolution forces geared up for battle.[2]

As the lines and tactics of the struggle shifted in the middle of the nation, Kansas became just one front in a war that seems to be being waged everywhere in America. Even the president of the United States stepped into the fray, casually recommending that schools teach students "both sides" of the debate.[3] Today nothing seems to be able to hold the battle back.

TEARING THE FABRIC

Late in 2001 the public television series *NOVA* aired eight hours of programming that explored the theory of evolution and the life of Charles Darwin. I served as an adviser to the series and even got the opportunity to appear on air in one of the episodes. As proud as I was of the series and its exceptional writers, producers, and actors, one sentence of narration in its final episode bothered me the first time I heard it. That last show, "What About God?," described a series of conflicts around the country regarding the teaching of evolution and then posed a question that surely was in the minds of many of its viewers:

> The majesty of our earth. The beauty of life. Are they the results of a natural process called evolution, or the work of a divine creator? This question is at the heart of a struggle that is threatening to tear our nation apart.[4]

Well, there was no question that it was a struggle, but "tear our nation apart"? *You've got to be kidding,* I thought to myself the first time I heard that narration. Slavery could (and almost did) tear this country apart; war or economic collapse might do so; but evolution? A couple of late-night squabbles at school board meetings might make for good discussion over morning coffee, but do they really have what it takes to rend the fabric of America? I didn't think so in 2001. But I certainly do now.

In Pennsylvania, Texas, Ohio, California, and Kansas, whenever the science curriculum comes up for public discussion, evolution becomes the key issue, the focus of controversy, and the turning point upon which communities become divided. The divisions, as many Americans know from firsthand experience, are deep and genuine. Those who defend science are regarded as godless atheists who wish the worst for our young people and seek to undermine both faith and traditional American values. When defenders of mainstream science strike back, they're often tempted to describe their tormentors as Luddites, fools, or worse, all because of their opposition to evolution. It's not a prospect that makes for reasoned discussion or pleasant discourse.

WHY HERE? WHY NOW?

"What the heck is going on over there?" a scientific friend from Britain asked me in 1999, just a few months after the Kansas Board of Education had voted evolution out of the science curriculum.

"What's the matter with you Yanks? Ya turnin' into a bunch of dummies and know-nothings?" Turning just a bit defensive, I got ready to play my trump card—six of the eight scientists who had won Nobel Prizes just a few months earlier were, in fact, Americans—but it turned out to be unnecessary, because my friend offered a solution: "Why, if this happened anywhere in Britain, we'd simply dispatch a couple of dons from Oxford or Cambridge." When these learned individuals arrived in the provinces, he went on, they'd lay out their university degrees and distinctions for the locals to admire. Then they would explain the standing that evolution held in the scientific community, in which they held positions of eminence and prestige, and that would be that. The local board would apologize, thank them kindly for their time, and then put its house back in order. Why wouldn't that work in the States? he wondered. I had a good laugh in response.

In America those eminent scientists would get the same three minutes of public comment as anybody else, and the state board, knowing damn well that its authority came from the voters and not from Harvard, Yale, or Kansas State, would proceed to do exactly what it wanted. It would toss evolution out of the curriculum, thumb its nose in the face of authority—scientific, academic, or otherwise—and be proud of what it had done. It's an American thing, I told my British friend. You just wouldn't understand.

While I do take a certain pride in the ornery nature of us former colonists, especially in conversations with folks from the mother country, that characteristic hardly provides an explanation for why this conflict has grown to such a scale in the United States, let alone why the conflict came into existence in the first place. So what *is* going on? Why is an issue that many scientists thought to be settled more than a century ago suddenly front and center in every corner of this great land?

In the pages that follow I will argue that evolution is only a placeholder in what is, in fact, a larger argument. Like an open plain where two armies have chosen to do battle, it has strategic value, but it's not really the point of the greater struggle. The stakes in the battle over evolution, as I will argue, concern much more than just Darwin's great idea. And we cannot resolve the issues at hand until we understand what the disagreement is really all about. To do otherwise would be like analyzing the Battle of Gettysburg in isolation from the causes of the Civil War itself. The struggles for Little Roundtop, the Peach Orchard, and Cemetery Ridge might have made perfect tactical sense in the context of this particular engagement, but they leave unanswered a sweeping, more fundamental question: What in the name of heaven were 150,000 soldiers doing in a small town in Pennsylvania, and why were they so willing to fight to the death over it?

Civil War historians, of course, have written volumes on the profound contradictions in American history and society that made the conflict possible, even inevitable. A thorough study of war, after all, is only partly about troop movements, weapons, and the possession of high ground. The deeper questions seek an understanding of what causes conflict, and how such causes play out in the ways in which wars are fought, won, and lost. To understand America's "civil war" over evolution, we have to examine remarkably similar questions about American science and culture. Is there something unique in the American character that bore the seeds of this conflict and provided fertile ground in which it could flourish? I think there is, and I'm not at all ashamed of that. In fact, I'm downright proud of it.

SCIENCE IN AMERICA

America is the greatest scientific nation in the world. It is true, of course, that America is not the only place where science is done, and done well. Scientific research is an international activity, one that transcends borders, cultures, and ideologies. Although modern science originated as a product of the European Enlightenment, deriving the core of its values from the drive to apply empirical rationality to understanding nature, it has long since spread far beyond that continent. As the United States rose to world prominence in the nineteenth century, it became a country uniquely hospitable to science, and science flourished here as never before.

American universities and research institutions lead the world in nearly every category of science, a fact borne out by statistics. Scientists working in America have dominated the Nobel Prizes in every field of science, and this dominance shows no signs of faltering. During the decade ending in 2004, for example, Americans won 71 percent of the Nobel Prizes in physics, 61 percent of the Nobels in medicine or physiology, and 58 percent of the chemistry prizes.[5] While many of these Nobel laureates were foreign-born and came to the United States to pursue their scientific careers, they did so for a compelling reason: America is simply where it's at for the practice of science.

Historians and sociologists have written volumes on the Americanization of international science, and they point to a variety of factors to account for it. Wealth, industrial might, our university system, and our relative freedom from war and social upheaval (especially when compared to Europe) have all played a part, but I believe there's a factor that transcends all of these: good, old-fashioned American rebelliousness.

Science, first and foremost, is a revolutionary activity. A genuinely new discovery changes our view of the world around us. Truly great science overturns our accepted ideas of nature, and therefore always presents a threat to the established order. To be effective in science a young investigator has to feel free to contradict and even to disrespect scientific authority. He or she has to be bold (or foolish) enough to do work that flies in the face of existing ideas, and then must be willing to set his or her career on the line to continue that research. Of course, remarkable discoveries don't always come from contradictions of canonical ideas—sometimes they emerge from work that simply goes off in a direction no one had ever anticipated. A young scientist finds a new frontier that has been previously overlooked or ignored, and it turns out to be a promised land instead of the dead end his mentors or supervisors might have feared. Either way, the most important ingredient for scientific success is the willingness of its practitioners to break free of the pack and try out their own ideas—to think, act, and work as true individuals.

My colleague at Brown University, the renowned historian Gordon Wood, has made a career of studying and analyzing the American Revolution. Wood's many insights into the formative years of our republic are extraordinary, and I cannot do them justice here.[6] One of his observations, however, has always resonated with me. In the course of little more than a decade, according to Wood, the Revolution shattered the intricate ties of obligation, deference, and respect that bound the prerevolutionary social order in colonial America. Americans changed from a people little different from the hierarchical societies of European monarchies to one that took up the truly radical notion that individuals were both the source of a government's legitimacy and its greatest hope for progress.

Out of this tradition grew public monuments like the extraordinary eleven-foot-tall statue that stands at the very top of the statehouse dome in Providence, Rhode Island, not far from my lab. Visitors to the state often mistake it for a duplicate of *Liberty,* the figure placed (several decades later) on the dome of the U.S. Capitol in Washington, but Rhode Islanders weren't thinking of liberty when their own capitol was built. This little state, site of the first violent act of the American Revolution,[7] commissioned a work known as *The Independent Man* to stand atop the seat of its government. The revolutionary symbolism is unequivocal: America is home to independent-minded individuals for whom a primary virtue is disrespect for authority.

And there you have it. Disrespect—that's the key. It's the reason that our country has embraced science so thoroughly, and why America has served as a beacon to scientists from all over the world. A healthy disrespect for authority is part of the American character, and it permeates our institutions, including the institution of science. For decades the United States was the one country in which, for a young scientist, getting ahead did not mean following the lead of a supervisor, a laboratory head, or a department chair.

American scientists, once they had earned their degrees, could apply directly, through a wide variety of competitive grant mechanisms, for money to support their work. The entrepreneurial nature of industrial research and development, part of the individualistic capitalism that grew out of the Revolution, fostered the same culture of disrespect, of initiative and free inquiry, and further stoked the fires of scientific creativity. European and Asian scientists flocked to America because of its great universities and research laboratories, of course, but also because of its invigorating climate of individualism. Scientists in this country, whether American by birth or choice, have been allowed to dream of revolutionary discoveries, and those dreams have come true more often in this country than in any other.

SCIENCE AND THE AMERICAN SOUL

If rebellion and disrespect are indeed part of the American scientific character, then what should we make of the antievolution movement? One part of the analysis is clear. The willingness of Americans to reject established authority has played a major role in the way that local activists have managed to push ideas such as scientific creationism and intelligent design into local schools. Education in America is fundamentally a local responsibility, despite efforts to guide and influence it on a state and national level. What this means, of course, is that individuals feel perfectly free to promote the inclusion of such ideas in their local schools, and so they have. At the state level antievolution movements have been able to introduce their own lesson plans, reword curriculum standards, and even redefine science itself. And all of this has happened despite clear and sometimes determined opposition from the "scientific establishment." In other countries citizens might have drawn back when university presidents and nationally recognized scientists said they were wrong—but not in America. In a democracy popular education inevitably reflects popular opinion, which, in the case of Darwin, narrowly supports the rejection of evolutionary ideas. This much, I would argue, should be clear and noncontroversial.

Now to the deeper question. Do the antievolution and intelligent design movements represent genuine revolutionary scientific ideas? Has American individualism once again given rise to groundbreaking science of the sort that would not be possible in a more authoritarian or hierarchical society? The advocates of these ideas certainly seem to believe they have accomplished precisely that, and portray the scientific establishment as the modern-day counterparts of those who refused to listen to Galileo, Einstein, Pasteur, and other scientific visionaries.

The "new science of design," its advocates claim, would give us an entirely new way to look at life. While evolution is limited to bringing about very slight changes, intelligent design (ID) would enable us to develop new antibiotics that would conquer disease once and for all. Accepting that tiny structures in the living cell were the product of intelligence would enable us to ask what the designer had in mind, and that might open new avenues of understanding as to how cells and tissues are put together. Small wonder that Michael Behe, one of ID's leading theorists, called "design" one of the greatest scientific discoveries of all time.

In this context today's conflicts can therefore be simply understood as the last desperate gasps of the dying theory of evolution, struggling against a new and vigorous idea that is about to replace it. This is exactly the point of view that ID's most passionate advocates project, and they fairly exult in their status as scientific outsiders. After all, they are quick to point out, radical new ideas such as quantum mechanics, continental drift, and even the big bang were once regarded as nonsense. And each of these is now an accepted part of the scientific mainstream. Why shouldn't ID be next?

Like any scientist, I have a built-in affection for the underdog, and like other researchers, I do not dream of conducting boring experiments that merely confirm the status quo, but of achieving new insights, radical and subversive, that would smash the orthodox and set off revolutions of thought and understanding. True scientists are not afraid of making waves. Rather, we dream of doing so, we lust for it, we fantasize about it.

As you have learned, dear reader, I am hardly a disinterested observer to the evolution wars. I have, in fact, been a partisan, an active and willing participant on the side of evolutionary science since the early 1980s. You might say that I am deeply invested in the cause. Yet none of this prevents me, in a certain sense, from imagining, almost hoping, that there might be something of value in the intelligent design movement.

I am a cell biologist, and my roots in that discipline go way back to middle school. After I had worked my way through the experiments in my Gilbert chemistry set and built the projects in my Erector set, I began pestering Santa for a microscope—and he delivered.

I was transfixed by cells from the very first time I looked through that simple device, certain that these little units, these tiny worlds of life, contained the greatest secrets in the universe. Today I am more certain of that than ever. The cell is literally what textbooks tell us it is—namely, the smallest unit of an organism that is truly alive. If that definition seems a bit ordinary, consider that the parts of any living cell (proteins, sugars, even nucleic acids) are just molecules—complicated molecules, to be sure, but just inanimate, dead molecules. Put them all together, and they spring into action; they become alive. And where, at what level, does this take place? At the level of the cell. The cell is the level of organization where life truly begins. The secret of life is in the cell if it is anywhere.

Imagine, therefore, the possibility that cell biology—the discipline in which I have published, taught, and worked for my entire career—might be in a position to revolutionize all of biology. Imagine that I and my fellow cell biologists might be able to tap our colleagues on the shoulders and tell them that we'd discovered something that would change all of their ideas about life, help them solve persistent scientific problems of the past, and lead them into radically new and productive directions. That would be pretty heady stuff—and, if traces of "design" could really be shown at the level of the cell, that's exactly what we'd be able to do.

So, the investigation I propose we undertake has two quite different goals. First, let's see if there is anything to ID. Let's examine the science behind the antievolution claims of the ID movement and determine whether we really are looking at the latest and greatest scientific revolution of our times.

If we are, we can stop right there, because the reasons for the popularity of the movement will have become obvious. ID is gaining support because it's correct and because it provides a new explanation of life that supersedes Darwinian explanations. A great conflict with scientific orthodoxy is to be expected as the movement disposes of evolution's rotting corpse, which helps explain the trials, the accusations, and the recriminations of those on the losing side of the argument.

But if this is not the case—if intelligent design is in fact a scientific impostor—then our inquiry must continue, because we have other questions to address. If ID is, as its critics claim, deeply and profoundly incorrect, why has this idea gained such popular support? Why are school boards around the country tempted to include it in their curricula, and why do a majority of the American people, including their president, insist that it should be taught alongside evolution? These are questions that matter, and they matter for reasons that extend well beyond the biology classroom and even past the issue of science education itself.

Science has prospered in this country because to a great degree its character matches the American character. In short, America has a scientific soul. We are practical, pragmatic, demanding. We want to see the evidence, and because we tend not to rely on authority, we want to see it for ourselves. We value the individual, and we lionize those who have gone against the grain to pursue a dream, to prove a point, to fight for an idea. We serve as an incubator of ideas, an engine of scientific creativity that has lifted the condition of mankind everywhere and opened new horizons of understanding from which the rest of the world can draw.

For more than a century America has occupied a position of scientific leadership and has gradually come to take it for granted. Although neither war nor economic depression nor political conflict has been able to threaten it, I now fear that that is about to change, for something has arisen that may indeed signal a change in our national character. That something is most visible in the debate over evolution, but it extends far beyond the teaching of a single subject in the curriculum of a single scientific discipline. It reveals a deep and profound split in the American psyche, an unease that threatens the way we think of ourselves as a people, the place we hold for science in our lives, and the way in which we will move into the twenty-first century.

What is at stake, I am convinced, is nothing less than America's scientific soul.

TWO

Eden's Draftsmen

AS AN AMERICAN PRESIDENT once famously came to understand, the "vision thing" matters, for it forms the glue that holds a society together. Indeed, "Where there is no vision, the people perish: but he that keepeth the law, happy is he." (Proverbs 29:18) Without the law, in its most general sense, society does not work. Without a system of rights and responsibilities, without a vision of what is true and fair, civilizations do indeed fail, and the people surely perish.

One might say that such a vision was behind the rise of the West, as European nations moved to assume world dominance and cultural preeminence in the seventeenth and eighteenth centuries. That rise included the industrial and scientific revolutions, which by now have transformed life on every corner of our planet. But in many ways the powerful transformations that emerged from these revolutions had the effect of unsettling the very certainties that made them possible. As humankind learned enough of nature to harness, control, and even to explain the natural world, in the eyes of many in the West this knowledge seemed to undermine the sense of purpose and harmony from which those revolutions had sprung.

The English poet Matthew Arnold captured exactly this sense in his great poem "Dover Beach." Its thirty-seven lines begin with a romantic description of calm and moonlit seas in the English Channel, hinting only briefly at an "eternal note of sadness" in the waves. But to Arnold, something more important than the waves was receding at Dover:

> *The Sea of Faith*
> *Was once, too, at the full, and round earth's shore*
> *Lay like the folds of a bright girdle furl'd.*
> *But now I only hear*
> *Its melancholy, long, withdrawing roar,*
> *Retreating, to the breath*
> *Of the night-wind, down the vast edges drear*
> *And naked shingles of the world.*

Faith has drawn back, and what has it left in its wake? A dreary, unadorned reality, a sad, grim world in all its unkempt misery. Lest anyone underestimate the darkness of his vision, Arnold drives the point home with some of the most memorable lines in all of English poetry:

> *. . . the world, which seems*
> *To lie before us like a land of dreams,*
> *So various, so beautiful, so new,*
> *Hath really neither joy, nor love, nor light,*
> *Nor certitude, nor peace, nor help for pain;*
> *And we are here as on a darkling plain*
> *Swept with confused alarms of struggle and flight,*
> *Where ignorant armies clash by night.*

Although Arnold did not publish "Dover Beach" until 1867, he actually wrote the poem in 1852, seven years before the publication of *On the Origin of Species* by Charles Darwin. The "melancholy, long, withdrawing roar" of the sea of faith had started, it seems, well before Darwin's work was to push it along. The poem, therefore, is not about the impact of evolution directly, but it certainly concerns the loss of vision, certainty, and faith. The "darkling plain" of Arnold's poem could surely bear the "dark satanic mills" that William Blake saw as the fruit of the industrial revolution, but just as surely it could describe a land of "nature red in tooth and claw," Tennyson's remarkable line that is now linked forever with Darwin's view of the struggle for existence. In ordinary terms neither of these is a happy vision, and neither will bind a people together.

To many, that's exactly the problem with evolution. If we are indeed the result of a process as natural as the weather, then our self-centered view of life would have to change. If we are here purely as a matter of luck, how could we be part of a greater plan? And isn't the vision of that greater plan essential to our unity as a people? Isn't it part of the legacy of faith that holds Western civilization together?

If Darwin's work has done anything, it would seem, it has broken those old ties to a common vision of a world in which we were part of the creator's providential plan for existence. It's broken *us*. We are alone, at night, on Dover beach, watching the sad tide of faith roll endlessly out to sea. This, it seems, is evolution's grim message for our times.

WRITING THE BOOK OF LIFE

Suppose, for a moment, that you were able to contact Matthew Arnold and tell him you had news for him, wonderful news. Deep inside the tissues of life, at the heart of the cell itself, you had found a message—not exactly a note in a bottle, but a message nonetheless. And it was a message that could have been written only by an intelligent agent. What that message would mean, as Arnold would grasp at once, is that there is an architect, a designer of life, and that the designer had us in mind as his greatest project. Suddenly the world of "Dover Beach," which had neither joy nor love nor certitude, would look very different indeed. It would dawn upon the poet in a flash that there was a kind of certitude afoot, and with that assurance it would be clear that our existence is no accident.

Well, there *are* messages in living things, and those messages are written in the language of a molecule called DNA. If Darwin's view of competition and struggle and opportunistic adaptation is cold and bleak, the discovery of a hidden message in life holds the warm and reassuring prospect of discovering the hidden plans of the author of that message. Not only would the message confirm his existence, it would confirm his direct involvement in every aspect of our lives. The tide of faith would surely roll back in, and the bright garment of dreams would once again enfold the earth.

The vision would return. The people would not perish, but endure.

The intelligent design movement makes exactly this promise. Using the language of molecular and cellular biology, it assures us that there is an intelligent author behind the complexity of life and that material forces alone are not sufficient to account for it. Life is good, and the sun is rising on the cliffs of Dover beach.

The idea that nature is "designed" is an old and honorable one. Aristotle's attempts to grapple with the character of existence led him to define four distinct kinds of causes that could be applied to any object, man-made or natural. A bowl, for example, has a material cause (it is made out of plastic), a formal cause (it has the form of a concave bowl), an efficient cause (it was made at a factory by a bowl maker), and a final cause (its purpose, or intent, which is to serve as a holder of food). The four causes are easy to identify in the case of manufactured or crafted items, like gloves or baseball bats, but if we attempt to apply them to the natural world, we may have a bit of a problem. What's the purpose, the final cause, for example, of a mosquito? Of bubonic plague? Of a dandelion? Aristotle asserted that such a cause existed, whether we could immediately identify it or not, and the modern opponents of evolution in turn assert that the ultimate answer to final cause is found in the person and intent of the designer.

The rise of Christianity established for its believers that the ultimate designer was God the creator. Thomas Aquinas, the greatest of Christian philosophers, made this argument explicit: Wherever complex design exists, there must have been a designer; nature is complex; therefore nature must have had an intelligent designer. This straightforward argument is one of Aquinas's five ways to demonstrate the existence of God, and was adapted brilliantly in Rev. William Paley's 1802 book, *Natural Theology.*

Paley opened his text with an image so powerful that it frames the popular debate on evolution even today:

> In crossing a heath, suppose I pitched my foot against a *stone,* and were asked how the stone came to be there; I might possibly answer, that, for any thing I knew to the contrary, it had lain there for ever: nor would it perhaps be very easy to show the absurdity of this answer. But suppose I had found a *watch* upon the ground, and it should be inquired how the watch happened to be in that place; I should hardly think of the answer which I had before given, that, for any thing I knew, the watch might have always been there. Yet why should not this answer serve for the watch as well as for the stone? why is it not as admissible in the second case, as in the first?[1]

In short, when we look at the complexity of a watch, we know something about it that we wouldn't dare conclude about the rock. Only a page or two later, Paley makes it clear what that conclusion is:

> The inference, we think, is inevitable, that the watch must have had a maker: that there must have existed, at some time, and at some place or other, an artificer or artificers who formed it for the purpose which we find it actually to answer; who comprehended its construction, and designed its use.[2]

The complexity of the watch implies that its parts were intentionally crafted and assembled for a purpose, a purpose that Aristotle himself would recognize as a final cause—namely, the purpose of telling time. If we can tell that about a watch, Paley reasons, why can't we do the same thing for the complicated organs and structures of living things? Are living things simpler than human devices, such as watches and telescopes? Not at all. In fact, they are far more complicated. If something as simple as a watch demands a designer, then surely the complex organs of living things do, too.

> For every indication of contrivance, every manifestation of design, which existed in the watch, exists in the works of nature; with the difference, on the side of nature, of being greater and more, and that in a degree which exceeds all computation.[3]

In his appreciation of the complexity of biological systems, Paley was exactly right. Living things *are* a lot more complex than nonliving things, a fact that remains true even in an age of extreme miniaturization and nanomachines. One might say, harking back to the Greeks, that Paley built his argument around Aristotle's third and fourth causes. Since no one had yet proposed a mechanism that might produce an eye or an ear or a beating heart, there were no immediately obvious efficient causes for any of these; yet since the final causes of these organs were obvious—to see, to hear, and to pump blood—there must have been a designer to determine those final causes and to construct their instruments accordingly. An intelligent designer, at that. The efficient causes of such organs still remained a mystery—or, at least, the trade secret of a creator for whom all things might be possible.

Paley's work was widely known in nineteenth-century England, and one of his most avid readers was a young naturalist named Charles Darwin. As he was later to record in his autobiography:

> In order to pass the B.A. examination, it was, also, necessary to get up Paley's *Evidences of Christianity,* and his

> *Moral Philosophy*. . . . The logic of this book and as I may add of his *Natural Theology* gave me as much delight as did Euclid. The careful study of these works, without attempting to learn any part by rote, was the only part of the Academical Course which, as I then felt and as I still believe, was of the least use to me in the education of my mind. I did not at that time trouble myself about Paley's premises; and taking these on trust I was charmed and convinced by the long line of argumentation.[4]

At one point, just before the publication of *On the Origin of Species,* Darwin wrote: "I do not think I hardly ever admired a book more than Paley's 'Natural Theology.' I could almost formerly have said it by heart."[5] However valuable he found Paley's thinking, the young Darwin would ultimately push it aside—literally and figuratively.

What Darwin did was to identify an efficient cause, natural selection, that shaped and molded living things. Once that efficient cause was in place, the final cause took care of itself. In fact, the final cause of every organism could be summed up in just one word—survival. Darwin, aided by his reading of Thomas Malthus, demonstrated that nature was dominated by what he called "the struggle for existence." Success in this struggle, measured by a species' ability to leave as many descendants as possible in the next generation, automatically adjusted its characteristics toward those that provided the best chance of survival and reproduction. From that perspective the purpose, or final cause, of organs like the heart or ear was not simply to hear or to pump blood; the purpose of these and every other organ, tissue, and structure of a living organism was to ensure that it prevailed in the struggle for existence.

Suddenly Aristotle's four causes were explicable by a theory built on forces that could be directly observed and verified, and the study of life no longer needed to infer purpose and invoke the miraculous as an efficient cause.

Paleyism was truly dead—or so the evolutionist might have thought.

But Paley lives on well into the twenty-first century in the hearts, souls, and rhetoric of the intelligent design movement. Its modern adherents generally agree that the Darwinian answer of "survival" could indeed answer the question of final cause, but they aim their sights on mechanism. Darwin, they maintain, never really provided a workable efficient cause, or mechanism, for any of Paley's great examples of design.

That claim routinely surprises scientists, who know that Darwin addressed Paley's contentions repeatedly in *On the Origin of Species*. In fact, one of his most famous passages dealt directly with Paley's arguments regarding the eye:

> To suppose that the eye, with all its inimitable contrivances for adjusting the focus to different distances, for admitting different amounts of light, and for the correction of spherical and chromatic aberration, could have been formed by natural selection, seems, I freely confess, absurd in the highest possible degree.

Opponents of evolution love to quote these lines out of their original context, because they seem to imply a Darwin who is coming clean, confessing the great inadequacies of his theory. In reality, he's doing nothing of the sort. Look at the next few lines, and you'll see that Darwin is confronting Paley's great challenge in the most direct way possible:

> Yet reason tells me, that if numerous gradations from a perfect and complex eye to one very imperfect and simple, each grade being useful to its possessor, can be shown to exist; if further, the eye does vary ever so slightly, and the variations be inherited, which is certainly the case; and if any variation or modification in the organ be ever useful to an animal under changing conditions of life, then

> the difficulty of believing that a perfect and complex eye could be formed by natural selection, though insuperable by our imagination, can hardly be considered real.[6]

By directly addressing the perfection of the eye, an argument whose pedigree predates even Paley by more than a century,[7] Darwin seemed to have taken on the most persuasive of all the design arguments and beaten it conclusively.[8]

But Darwin had sought only to show that an evolutionary pathway was possible, not to delineate a detailed step-by-step process that had actually given rise to the eye as we know it today. In this respect his explanation was incomplete, which is hardly surprising, given that scientific explanations are always, to some extent, incomplete. For all of their apparent precision, even gravitational theory and atomic theory are incomplete in their descriptions of the mechanics of nature. Working in a time before genetics, before biochemistry, and at the infancy of developmental biology, Darwin did not and could not account for the workings of each step in the pathway. What that means, to those who view evolution with suspicion and dread, is that Darwin didn't really demolish Paley. The argument from design still stands; indeed, in their view, it may yet prevail.

So why does scientific history conclude otherwise? Biochemist Michael Behe, one of the leaders in the ID movement, argues that Paley's legacy suffers from his failure to identify the general principle that makes organs such as the eye unevolvable:

> Almost none of his examples has been specifically refuted by demonstrating that the features could arise without a designer, but because for many examples Paley appeals to no principle that would prevent incremental development, people have assumed since Darwin that such gradual development is possible.[9]

According to Behe, however, there is indeed such a principle, one that makes gradual development of the kind envisioned by

Darwin literally impossible. That principle is "irreducible complexity," the claim that complex biological systems are composed of multiple parts, and that the removal of just one part would effectively cause the system to stop functioning. If the eye, the ear, the heart are irreducibly complex, then they couldn't have evolved, because those intermediate "gradations" would be missing key parts, and in that form wouldn't work. Without some useful function to give those bits and pieces survival value, natural selection could never drive the pathway leading to the final organ, and evolution would lose its efficient cause.

The great discovery of today's intelligent design movement, according to its proponents, is that irreducibly complex systems do exist, and that they are vital to life itself. Where do we find these systems? In a place where Darwin and the HMS *Beagle* never dropped anchor: deep inside the living cell.

A UNIVERSE IN MINIATURE

I teach a course in cell biology every fall at Brown University. It's an upper-level course, and because nearly all of the reading I require of my students comes from the original scientific literature, they have to read and analyze it directly. For several years I've begun the semester by assigning a paper from the journal *Cell* written by Bruce Alberts, a distinguished cell biologist who served for seven years as president of the National Academy of Sciences, probably the world's most prestigious scientific group. In "The Cell as a Collection of Protein Machines," which was intended as advice to science educators like me, Alberts wrote:

> We now know that nearly every major process in a cell is carried out by assemblies of 10 or more protein molecules. And, as it carries out its biological functions, each of these protein assemblies interacts with several other large complexes of proteins. Indeed, the entire cell can be viewed as a factory that contains an elaborate network of

interlocking assembly lines, each of which is composed of a set of large protein machines.[10]

What this means, of course, is that to prepare a new generation of scientists ready to grapple with this sort of complexity, we have to make sure that our students learn the principles of engineering and manufacturing that apply to these "machines" within the cell. This is sound educational and scientific guidance, and it's fair to say that I devote a majority of classroom time in my cell biology course to exploring, explaining, and understanding these machines.

To reinforce this point one could observe that the day-to-day shoptalk of biology today often mentions "motors" that generate force, "gatekeepers" that control entry and exit to parts of the cell, "pumps" that move molecules between compartments, and even "checkpoints" that carefully monitor the conditions within the cell before allowing a new process to begin. All of these machines are complex, and all of them are composed of multiple parts.

So pervasive is this mode of thinking that I even use it as my way of explaining the living cell to younger students. In one of the textbooks that Joe Levine and I have written for high school biology, we compare the cell to a complex modern factory, and make analogies between the parts of the cell and the parts of a manufacturing plant. The nucleus is the factory's main office, the mitochondria its power plants, the ribosomes its manufacturing equipment, and the Golgi apparatus its shipping and receiving department. All teachers use analogies to get across complex material, and in writing a textbook we do everything we can to assist students and teachers with examples like these. They work because they are easy to remember, and because they make scientific sense.

If the parts of a cell really are like the complex components of a factory, what does that say about evolution? Let's start with another analogy. If the cell is like a factory, we might ask where factories come from and then apply that answer to the cell. Factories, of course, are human inventions. They are built, staffed, and designed by human beings and are in every sense the product of human intelligence. Extending that analogy, therefore, would suggest that cells, too, must be the product of intelligence, although it does not tell us what kind of intelligence, since cells are truly quite different from factories. One might put it another way: By analogy, cells must be "designed" in the same way that factories are designed.

If we examine a cell and identify a protein that functions as a molecular pump, and we know that water pumps and fuel pumps have intelligent designers, then why wouldn't a cellular pump also have a designer, a designer with the intelligence to create it and the foresight to leave the plans for its construction in the coded language of DNA?

THE MOUSETRAP OF DESIGN

It should be obvious, however, that we cannot just point to cellular complexity and shout, "Evolve that!" at the pro-Darwin crowd. As they would quickly point out, the fact that evolution has not yet explained how a particular structure or pathway or molecular machine came to be is no guarantee that it cannot ultimately be explained by evolution. They might be bold enough to remind us that even a phenomenon as commonplace as the shining of the sun was thought to be a violation of the laws of nature until nuclear fusion was discovered and physicists realized that the presence of solar hydrogen and helium were definitive signatures of the fusion process. Taking what is unknown, unexplained, or undiscovered today and claiming that it will remain forever beyond our understanding isn't just poor logic—it's a lousy bet, considering the rate at which science continues to advance.

No, we need something more definite than mere skepticism built around complexity to make a compelling case for the intentional and intelligent design of life. We must be able to put forward an explanation that tells us exactly why evolution will never be able to account for the machinery of the cell. We must find a reason why such machines are literally unevolvable.

We can find that reason by going back to our analogy. Let's take a very simple multipart machine, such as a mousetrap. A mousetrap has five parts, and as Michael Behe has pointed out, you need all the parts of the mousetrap to make it function properly:

> The mousetrap depends critically on the presence of all five of its components; if there were no spring, the mouse would not be pinned to the base; if there were no platform, the other pieces would fall apart; and so on. The function of the mousetrap requires all the pieces: you cannot catch a few mice with just a platform, add a spring and catch a few more mice, add a holding bar and catch a few more. All of the components have to be in place before any mice are caught.[11]

No one, of course, has ever claimed that mousetraps were the products of Darwinian evolution, but that's not the reason for Behe's interest. Mousetraps are *machines,* simple ones, but machines nonetheless, and they are composed of multiple parts, just like the more complex machines inside the living cell. It's easy to see how the absence of even just one part can destroy the function of the mousetrap. You can't take one of its five components away and still have 80 percent of mousetrapping function left. Remove even one part, and the mice will run free. In reality you've got zero functionality if just one element is missing. That's the salient point, and that's why mousetraps are relevant to evolution.

In the language of intelligent design, a mousetrap is "irreducibly complex," and there you have it—that's the argument that Paley missed, that's the cornerstone of the new, scientific, intelligent design movement. The structures and organs of the body, as Paley realized, are complex. So are the machines of the cell, which he could not appreciate in his lifetime. But what Paley should have added to his argument, the argument from design, was that their complexity is "irreducible." Take a part away, and useful function disappears.

Remember Charles Darwin's point about "numerous gradations…imperfect and simple"? Well, if none of those numerous gradations worked because they were all missing a part or two, then natural selection would have found them useless and could never have incorporated them into the gradual pathways leading to the evolution of complex machines in the cell. What this means is that, to the extent that the machinery of life can be compared to a mousetrap, evolution is wrong. The question then becomes whether we can find irreducibly complex machines inside the cell. And the great news, from the point of intelligent design, is that we can. In fact, they are everywhere.

RUBE GOLDBERG IN THE BLOOD

One of the hallmarks of an irreducibly complex system is that it should be brittle and easy to break. If any one part is missing, its absence might even be life-threatening.

Striking examples of exactly such systems have been known for centuries. In 1803 an American physician named John Conrad Otto published a paper entitled "An Account of an Hemorrhagic Disposition in Certain Families." Dr. Otto described three generations of what was then known as "the bleeder's disease" in the male descendants of a woman who had lived in Plymouth, New Hampshire, in the early 1700s. Otto was not the first person to take note of this curious disease. Talmudic law, dating from the second century A.D., recommended that male babies not be circumcised if two of the baby's brothers had already died from bleeding after the procedure. That religious exemption recognized that hemophilia, as the bleeder's disease is now known, ran in families and was far more common in males than in females. In today's terminology, it was a sex-linked genetic disorder. The blood of hemophiliacs is not able to clot properly, and left untreated they are at risk from death due to uncontrolled bleeding and internal hemorrhage.

Biochemical and medical studies in the 1940s showed that there were actually two forms of the disorder (now called hemophilia A and B), and twenty years later the exact proteins responsible for each form were determined. By the beginning of the 1970s the causes of the disorder were understood to be deficiencies in one of two factors in a complicated, interacting network of proteins known as the blood-clotting cascade (figure 2.1). Hemophilia A is caused by a deficiency in a protein known as factor VIII, and hemophilia B by a lack of factor IX. If you are missing either one, your blood doesn't clot properly, and your life is at risk.

The overall pathway of clotting is complex enough to torture college biochemistry majors, and some of my colleagues use it (intentionally or not) for exactly that purpose. As much as anything in biology, it frankly resembles a Rube Goldberg machine, one of those lovable contraptions drawn by the great cartoonist to carry out very simple tasks, such as switching on a light or opening a napkin, by means of exceedingly complex machines built out of an improbably large number of strange parts. Could a cellular equivalent of a Rube Goldberg machine have been produced by evolution? It certainly doesn't seem possible. Because each and every step in such a machine is unique, and the loss of just one part would abolish its function, it's hard to imagine how it could have been assembled one or two steps at a time, as evolution would seem to require.

We celebrate Rube Goldberg precisely because of his genius in designing ever more complex ways to make us laugh. Whoever designed the blood-clotting system, however, had something different in mind. By placing the final event in the formation of a clot—the conversion of a soluble protein called fibrinogen into an insoluble fiberlike protein called fibrin—at the end of a long cascade of steps, the designer made it possible to take a tiny signal and amplify it over and over again. Even a minuscule break in a blood vessel can trigger the cascade, which produces a stronger and stronger signal at each step, until the final reaction produces enough fibrin to seal the break and prevent damage to the circulatory system. But it doesn't stop there. The multistep system also includes a series of checks and balances ensuring that the clotting reaction doesn't get out of hand, controlling the amount of fibrin

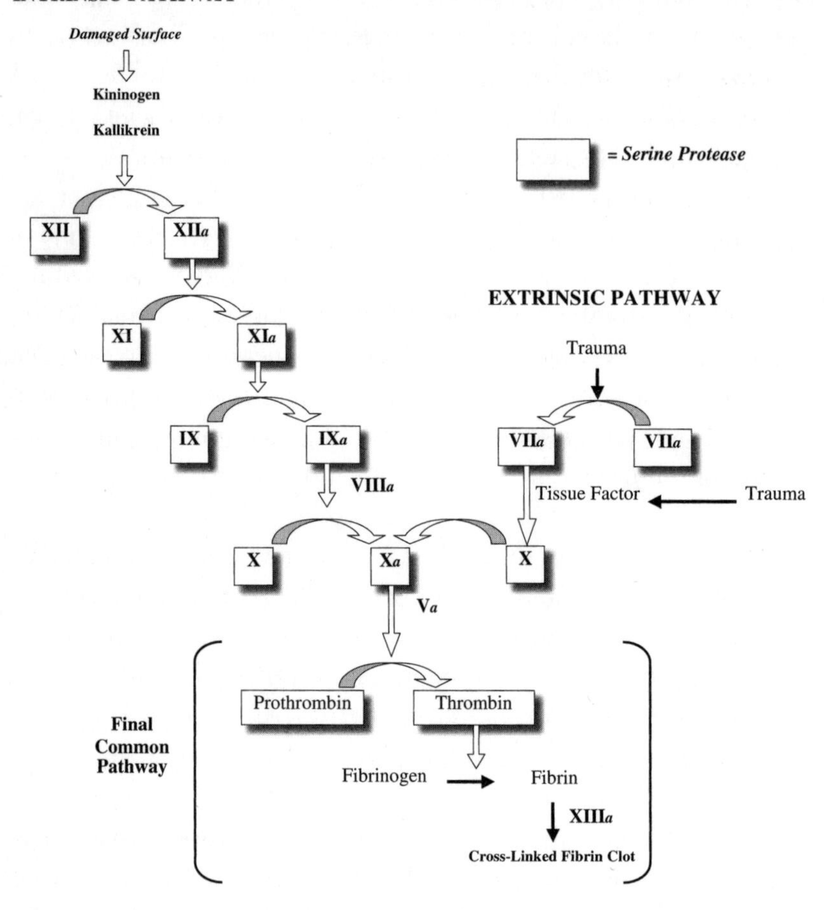

Figure 2.1: The blood-clotting cascade. This diagram shows the major components of the human blood-clotting pathway. Each component of the pathway represents a "clotting factor," in most cases a protein, the activation of which triggers the next step. The horizontal arrows represent conversion of a factor from its inactive form to the active form, and the vertical arrows indicate triggering of the next step. Boxes have been drawn around the factors that belong to the serine protease family of enzymes. The final result of the pathway, shown at the bottom, is the formation of a clot of cross-linked fibrin proteins that stops bleeding. Intelligent design argues that the pathway cannot work until all of these factors are in place, so it could not have been produced by a gradual, step-by-step process like evolution. *(Kenneth R. Miller.)*

produced, and restraining the clotting reaction just enough to keep blood flowing elsewhere in the body.

We might say, as Paley would have, that the elegance and complexity of the system compels us to invoke a designer. But today, by invoking intelligent design, we can go further.

Remember what happens in hemophilia? The lack of just one component produces a potentially fatal defect and requires constant medical attention. What this seems to show is that the human blood-clotting cascade is also a perfect example of an irreducibly complex system, a system that must have been designed. Can there be any doubt about this? Not according to Michael Behe, who has held up blood clotting as one of the prime examples of scientific evidence for intelligent design. Behe discussed the clotting system in a chapter he helped to write for the widely distributed intelligent design textbook *Of Pandas and People:*

> However, biochemical investigation has shown that blood clotting is a very complex, intricately-woven system containing a score of interdependent protein parts. The absence or defective operation of any of several of these components will cause the system to fail, and blood will not clot at the proper time or at the proper place.[12]

Consider what this means: If each and every part of the system has to be present simultaneously for blood to clot, then the system could never have been produced by gradual, step-by-step evolution. It is indeed irreducibly complex and therefore unevolvable. If Darwinian evolution could not have produced it, what could? The answer, of course, must be design.

Behe made this point even more unequivocally in his popular book on intelligent design, *Darwin's Black Box:*

> Since each step necessarily requires several parts, not only is the entire blood-clotting system irreducibly complex, but so is each step in the pathway.[13]

> ... In the absence of any of the components, blood does not clot, and the system fails.[14]

Darwinian evolution, in the words of Darwin himself, requires "numerous gradations" on the way to a complex system, and every one of them has to be advantageous: They all have to work. The irreducible complexity of blood clotting, however, shows that the absence of even a single part of the pathway would be fatal. Find as many fossils as you like, one might say, but it doesn't matter if evolution cannot clot the blood. These marvelous little Rube Goldberg machines floating around inside us are surely the handiwork of the designer, and that's all the proof we need.

SPINNING THE FLAGELLUM

If proof of design in your blood isn't enough, how about in your stomach? Our digestive systems play host to billions of bacteria, most of them happily swimming around, reproducing, and even helping us to digest our food. Many of them, including the famous *Escherichia coli (E. coli),* zip along powered by nifty little motors that spin tiny whiplike flagella (figure 2.2).

Make no mistake about it, these flagella are marvelous little machines. Powered by hydrogen ions, they are high-efficiency, reversible rotary engines that respond to internal and external signals and can maneuver cells with extraordinary speed and dexterity. Although all of the details of their function are not completely understood, we certainly know enough to stand in awe of their construction. David DeRosier, a structural biologist at Brandeis University, has aptly written that "more so than other motors, the flagellum resembles a machine designed by a human."[15] *"Designed by a human"?* How much clearer could it be that the flagellum must be the product of an intelligence?

The flagellum has been used so often as an example of intelligent design that one might almost call it the poster child of the movement. It shows up regularly on intelligent design Web sites, in

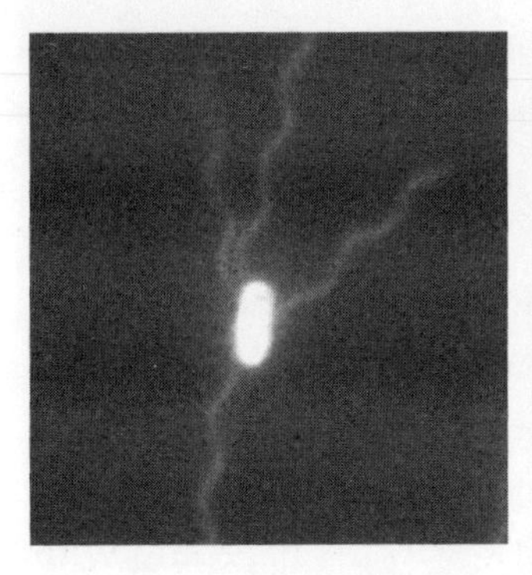

Figure 2.2: The bacterial flagellum. A fluorescence microscope image of *Escherichia coli (left)* shows four flagella extending from the cell. Powered by an ionic gradient across the cell membrane, these flagella rotate in a corkscrewlike fashion at high speed to propel the bacterium through liquid. The rotary "motor" that powers this movement *(right)* is an intricate biochemical machine built into the base of the flagellum. Consisting of thirty or more individual proteins, the complexity of the flagellum has made it a favorite for those who insist that evolution cannot account for such structures within a living cell. *(*Left: *Professor Howard Berg, Harvard University.* Right: *WGBH Television, from the* NOVA *science series.)*

ID articles, on book covers, and even makes appearances in court. In fact during the 2005 intelligent design trial in Pennsylvania, it was brought up so often that Scott Minnich, an ID proponent from Idaho State University, was forced to make a joke about it as he took the stand late in the trial. When the judge sighed, "We've seen that before," after being shown yet another image of the flagellum, Minnich deadpanned, "I kinda feel like Zsa Zsa [Gabor]'s fifth husband. I know what to do, but I don't know how to make it exciting."[16]

The inner workings of the flagellum are quite complex, and scientists will freely admit that they do not know how it evolved. That's understandable, since the flagellum is quite likely an ancient structure and may have been around for as many as a billion years. Nonetheless one could use it as a classic "argument from ignorance" against evolution. Since we don't know how it evolved, perhaps it was specially or miraculously created or designed.

But the value of the flagellum to intelligent design rises far above this simple contention and rests on its status as yet another concrete example of irreducible complexity. By asserting that it, too, is a system "in which the removal of an element would cause the whole system to cease functioning,"[17] the flagellum is presented as a "molecular machine" whose individual parts must have been specifically crafted to work as a unified assembly.

In the case of the flagellum, the assertion of irreducible complexity means that a minimum number of protein components—perhaps thirty—are required to produce a working flagellum with biological function. By the logic of irreducible complexity, these individual components should have no function whatsoever until all thirty are put into place, at which point motility appears. The flagellum starts spinning, and the mechanism suddenly becomes useful to the cell that possesses it. What this means, of course, is that evolution could not have fashioned those components a few parts at a time, since those parts do not have functions that could be favored by natural selection. As Behe wrote, "Natural selection can only choose among systems that are already working."[18]

Surely the flagellum proves the case for design. Its proponents may overuse it as an example, but overuse can be forgiven if the example is valid. And the flagellum certainly seems like the real deal, doesn't it?

CODE WRITERS

"Today, we are learning the language in which God created life. . . . We are gaining ever more awe for the complexity, the beauty, the wonder of God's most divine and sacred gift." Those were the words spoken by President Bill Clinton on June 26, 2000, when he gathered researchers at the White House to announce the successful completion of the Human Genome Project. Clinton's words were echoed by Francis Collins, the head of the federal genome effort: "It is awe-inspiring to realize that we have caught the first look at our own instruction book, previously known only to God."[19]

The human genome, the complete set of all our genes written in the language of DNA, is an impressive piece of engineering, even overwhelming in terms of its size and complexity. DNA encodes information in a four-letter alphabet, known as nucleotide bases, usually represented by the letters A, C, G, and T. Our genome contains more than three billion of these bases, so it's easy to imagine it as a string of one-letter characters typed on page after page of a very large book. How large a book? Well, when I type out a page of single-spaced text on my computer, it contains about three thousand letters. If the human genome were printed out in the same typeface that I use, it would produce a book of one million pages, or a thousand books of a thousand pages each.

What's written on those pages? To a surprising extent the answer seems to be "nothing." Only about 2 percent of the human genome seems to code for the essential business of life, while whole sections, tens of thousands of bases in length, seem to have no biological function at all. Molecular biologists actually call some of these regions "gene deserts," reflecting their barren nature. But in the nondesert regions—the oases, if you will—the genome is packed with the stuff of life. In active genes the sequences of bases in DNA are copied, or transcribed, into a similar molecule known as RNA.[20] Some RNA molecules have important functions of their own, but others direct the order in which amino acids, the building blocks of proteins, are strung together. By determining the sequences of amino acids, genes determine the ultimate properties of proteins, which do everything from digesting our food to producing movement in our muscles to replicating DNA itself in preparation for cell division.

If a modern William Paley needed an example of complexity with which to demonstrate the exquisite design of life, the genome would surely serve. The best current estimates are that these millions of pages contain somewhere between 25,000 and 30,000 genes, the coding units that produce a single protein or RNA molecule. Paley might have asked if anyone could seriously argue that such complexity could have arisen by chance. They couldn't, of course, and evolutionists would agree. Natural selection, they'd point out, is not a chance, random process at all, and they'd be right. Although Paley didn't know about natural selection, today we do and can therefore put the challenge to evolution somewhat differently.

The genome is chock-full of information. In fact, if we simply took those pages full of letters, those sequences of A's, G's, C's, and T's, and ran them through a computer program designed to detect information in coded messages, it would immediately identify many of the coding regions of our DNA. Those letters appear in certain patterns, and the existence of such patterns would reveal the role that DNA plays as a carrier of genetic information. So, the issue before us can be put in terms of information—does the mechanism of evolution have a way to generate the biological information found in our own DNA? Advocates of intelligent design think not.

One of them is Stephen Meyer, the director of the Discovery Institute's Center for Science and Culture, a pro-ID think tank based in Seattle. As Meyer has noted, although we share many genes with other organisms, one of the most important requirements for any workable theory of evolution is that it must include a mechanism capable of producing new genes with novel functions. In a 2004 paper he noted:

> Many scientists and mathematicians have questioned the ability of mutation and selection to generate information in the form of novel genes and proteins. Such skepticism often derives from consideration of the extreme improbability (and specificity) of functional genes and proteins.[21]

Meyer's point is that the sequences of amino acids (or the DNA bases that code for them) that can produce functional proteins are few and far between. In fact, they are so few that the evolutionary mechanisms he cites—namely, random mutations or changes in DNA—would never be able to produce them in any reasonable length of time. Therefore, something else has to be invoked to explain DNA's remarkable information content. That something, of course, is the hand of the designer.

William Dembski, a philosopher and mathematician also associated with the Discovery Institute, has worked to formalize this remarkable idea. According to Dembski, living systems contain what he calls "complex specified information" (CSI). CSI is actually commonplace in our lives, and we encounter it all the time. The telephone book, a newspaper, the bits and bleeps of a modem's signals between computers, and even your sixteen-digit credit card numbers are all examples of CSI. It shouldn't surprise us, therefore, that the DNA sequences in our genomes, which surpass any of these everyday phenomena in their complexity, are examples of CSI, too.

Where does CSI come from? In the natural world there are two general kinds of originators—natural ones and intelligent ones. Phone books and credit card numbers, like all human examples of CSI, are ultimately generated by the intelligence of human beings. The key question, however, is whether natural, unintelligent causes can also produce information of such complexity. This has been the focus of Dembski's work, and in one of his books, he presents the conclusions of his analyses:

> We can summarize our findings to this point: (1) Chance generates contingency, but not complex specified information. (2) Laws...generate neither contingency nor information, much less complex specified information. (3) Laws at best transmit already present information or else lose it. Given these findings, it seems intuitively obvious that no chance-law combination is going to generate information either.[22]

In other words, blind, undirected natural processes cannot generate complex specified information, and that includes the CSI found in living things. Dembski's mathematical analysis of information and its production led him even further, to the assertion that he has discovered a fundamental law of nature:

> *Natural causes are incapable of generating CSI.* I call this result the Law of Conservation of Information, or LCI for short.[23]

If there is indeed a law that does not allow for the production of CSI by natural causes, is evolution an exception to that law? Not according to Dembski:

> This argument holds for Darwin's mutation-selection mechanism, for genetic algorithms and indeed for any other chance-law combination.[24]

For Darwinian evolution the discovery and confirmation of a principle like the Law of Conservation of Information should be devastating. If living systems arose the way that evolution tells us they did, they must have generated new forms of information over and over again. Evolution requires the production of new genes, new proteins and biochemical pathways, and even new body structures. Each of these contain CSI—and if evolution, a natural process, cannot generate that CSI, then what did? The answer, of course, could be an intelligent designer who wrote that information into the coded language of DNA. In short, life contains information that could only have come from an intelligent source. Who really did write the book of life? Perhaps our own DNA holds the answer: It was an intelligent designer.

EDEN'S SECOND DRAFT

When "scientific" creationists attempted to challenge evolution in the 1970s and 1980s, their playbook was the Book of Genesis. Membership in any number of creationist organizations required the applicants to sign a statement attesting to the literal, historical truth of Scripture. A large chunk of their time and effort was devoted to defending the biblical account of creation, and even today creationists lead boat trips down the Grand Canyon of the Colorado, dismissing the towering testimony of distinct geological ages that surround them as the illusory product of a single worldwide flood. If the account of Eden was literally true, then creation was nearly simultaneous, with every plant and animal, living or extinct, churned out in just a few furious days of activity roughly six thousand years ago. The fossil record, with its detailed evidence of succession, extinction, and appearance, was a problem to be analyzed and solved with "flood geology," "ecological zonation," or any of a host of other desperate ideas purporting to explain why that record shouldn't be taken at face value.

In the early 1980s I debated two of the most prominent "young earth" creationists, Duane Gish and Henry Morris, a number of times. Their defense of Genesis and a six-thousand-year age for the earth dominated those debates, which ultimately set my creationist opponents back on their heels time after time, when they would rather have simply been taking potshots at Charles Darwin. As a debater there are few things I enjoy quite so much as facing an opponent who is willing to reject physics, geology, and astronomy along with biology. It was easy to show that Gish and Morris's real problem was with *all* of science, not just with Charles Darwin, and the audience quickly realized that Gish and Morris had actually rejected science itself in order to accommodate their biblical beliefs. In a practical country like the United States, that's not a stance that's guaranteed to win many converts, but the public relations problem it presents is one today's ID movement has found a way to circumvent.

They've done so not by changing their ultimate goals, but by writing a brand-new playbook, one that's lighter and infinitely more flexible. With a wink and a nod to the Bible, they've set that heavy book aside and stepped into the ring unencumbered by its literalist baggage. ID, they maintain, is a scientific theory, not a religious conviction, and therefore the age of the earth doesn't matter. By positing a designer as the source of each and every "evolutionary" novelty, design doesn't have to struggle with a worldwide flood, and need not take on earth scientists regarding the geologic column. That enables it to dismiss the geologic evidence of change over time with an indifferent shrug while continuing to snap off jabs at evolution.

Eden's second draft doesn't state when or where or how, so it can't be trapped on such points. It doesn't make the scientific claims of young-earth creationism, and that's its beauty as a weapon against Darwinism. It makes for such a deliciously small target that it suddenly becomes almost impossible to attack, and enables the adherents of ID to remain on the offensive. The evolutionist can then be forced to account for the information content of DNA, for the elegant complexity of the bacterial flagellum, for the missing details of the fossil record, and for the detailed, step-by-step evolution of each and every biochemical system in nature.

Evolutionists often try to fight back by pointing to the imperfections of biological design, examples of which are easy to come by. For anyone in late middle age with a bad back, the imperfect design of the human body is almost too obvious. Indeed Darwin himself thought the sort of "design" he saw in nature was unworthy of a kind and clever God. In an 1856 letter to Joseph Hooker, he wrote, "What a book a devil's chaplain might write on the clumsy, wasteful, blundering, low, and horridly cruel works of nature." But such criticism is easily answered by today's design movement, which insists that the detection of design does not require that that design be perfect—or even very good. Consider the Edsel or the Yugo or any other poorly designed piece of automotive machinery. However bad the results, there is no doubt that these cars were, in fact, designed—maybe not very well, but designed nonetheless. Imperfection, therefore, does not argue for evolution. It might tell us something about the talents of the designer, of course, but it certainly doesn't imply that he doesn't exist.

What we are left with, in the eyes of many Americans, is a compelling case for intelligent design. Evolution, after all, hasn't solved every problem in biology. Mysteries remain, and ID tells us that there is a reason why those stubborn questions will never be solved by Darwinian logic. They *can't* be. Darwinian evolution will always fall short, because the key features of living things were not produced by evolution. They are the products of design, and only a science that admits design as a genuine possibility will be able to take account of them and lead us into the scientific future.

If we can look for coded information in radio and telephone signals, we can certainly look for it in the genome. And once we find it, the case for design is set. The power of this argument is that every advance in our understanding of biology only seems to strengthen it. Learning more about our cells only increases our appreciation of just how complex they are, a complexity that surely calls out for a designer who put it all together.

The popular power of the ID argument rests on the ease with which one may point to any example of complexity and ask, "How come?" To prepare a cogent response, especially one that actually explains the evolution of complexity in an understandable manner, is beyond the popularizing ability of most scientists. And that means that the attacks of the ID movement place scientists and the scientific establishment in the pitiful defensive position of saying, "Trust us," on such issues.

And there's the problem: For many Americans the response is, "We just can't." It's for that reason, more than any other, that design is winning.

THREE

Embracing Design

THE MOST SINCERE COMPLIMENT anyone can pay to a scientific idea is to take it seriously. If such an idea is genuinely provocative, like continental drift, cold fusion, or general relativity, it will generate a buzz, a kind of creative excitement, in the scientific community. That sort of buzz will inspire both positive and negative reactions. Some people are going to hate the new idea, and some people are going to love it, but if it's novel and genuine, it's going to provoke a creative response. The precise nature of that response tends to surprise many people, especially those who are accustomed to thinking that scientists set out to "prove" an idea right or wrong. What happens much more often is that in their excitement to explore a novel possibility, they simply try to follow it, to see where it leads. They *embrace* it.

Cold fusion was a typical example. The announcement by Stanley Pons and Martin Fleischmann in 1989 that they had achieved nuclear fusion in a test tube at room temperature stunned the scientific community. Physicists everywhere rushed to set up similar experiments in their own labs, seeing the promise of a future of cheap energy—and maybe a slice of personal fame. Nuclear fusion, after all, was thought at the time to be possible only at extremes of temperature and pressure close to those inside the sun, and duplicating those conditions on earth was clearly a job for "big science"—big machines, big buildings, big budgets. The notion that it might be done under benchtop conditions was more than revolutionary; it was (almost literally) electrifying on a personal level, since physicists might be able to achieve the process at a scale within the reach of anyone's laboratory. In short, physicists everywhere embraced cold fusion.

In one lab after another, however, experiments simply didn't produce the results that Pons and Fleischmann had claimed. Patiently, carefully, researchers everywhere searched their protocols to see where they could have gone wrong. Slowly, gradually, they came to the conclusion that their own problems in achieving cold fusion were indicative of a fundamental reality—namely, that something was wrong with the concept itself. Today cold fusion (at least the Pons and Fleischmann variety) is a dead issue, a case study of how the enthusiastic embrace of science can kill a bad idea. Following this particular one to its logical conclusion led nowhere—except to the scrap heap of junk science.

Intelligent design, one would hope, has a lot more in common with continental drift or the big bang than it has with cold fusion. But there's only one sure way to find out if this is indeed the case, and that's to take the idea seriously. If design really is the key to the complexity of living things, we ought to be eager to do a lot more than merely say, "Well, that means there must be a designer." We should be embracing design as a central principle of earth history, biological development, genetics, and genome organization. Design should be the tool we use to understand fossils, to develop new drugs, and to classify organisms. After all, that's how we use evolution now, and if design is to replace Darwinian evolution, it should do all of those things—and it should do them a whole lot better.

A NATURAL HISTORY OF DESIGN

Evolutionists have never been shy about applying their theory to natural history. In fact, our expanding understanding of the natural history of Planet Earth played a major role in giving rise to evolutionary theory in the first place. If life on earth had remained unchanged over time, there would be no need for science to address the subject. But life has changed over the course of millions of years, and the changes have been dramatic.

These changes have also occurred in patterns, and since the effort to explain them has driven and shaped evolutionary theory, it should also shape design. What sort of patterns are these? First, there is a sequential character to the appearance of life on earth. The earliest organisms to appear in the fossil record are single-celled bacteria and algae, and the story of their existence actually occupies most of the history of life on earth. If those organisms were indeed designed, then the designer saw no need for anything much more complicated for quite a while. Only after nearly two billion years of life did the first cells with nuclei, which we know as eukaryotic cells, appear.

A few hundred million years later came the first multicellular organisms, although most of these were a far cry from the complex animals we know today. Then, in a remarkable 30-million-year period known as the Cambrian, a wide variety of animals appeared, including the ancestors of nearly all the major animal groups, or phyla, that we recognize today. As dramatic as this period was, however, the development of life didn't end when the Cambrian concluded 530 million years ago. After all, it had not given rise to insects, reptiles, or birds, or produced pine trees or flowering plants; dry land, at the conclusion of the Cambrian, was still devoid of animals and large flora, all of which were to come in the hundreds of millions of years to follow. Any theory of intelligent design must take these sequences into account. To put it plainly, if design is the explanation for life, then the designer has chosen to work gradually, assembling a living world over the course of hundreds of millions of years.

The second pattern in the fossil record is unmistakable to anyone who examines its individual species in detail. That pattern is revealed in the appearances of new organisms, and clearly suggests that new species are the descendants of ancient ones that preceded them. Now that may not be true, of course, since fossils are just fossils, and they don't tell us directly who their ancestors might be, or even if they really have them. It does mean, however, that if those ancestor-descendant relationships aren't real, then we need to explain how our intelligent designer managed to produce so many apparent family trees. We can see these patterns when we look for the earliest representatives of major animal groups, and they are striking. The first amphibians, for example, appeared on earth in the Devonian period, roughly 380 million years ago. They can indeed be recognized as amphibians, but their differences from modern amphibians are obvious. Not only do they contain internal gills (similar to fish) and a fishlike arrangement of bones in their skulls, but they also sport limbs with eight digits, the exact number of fingerlike projections seen in certain groups of fish in the geological period that preceded them. In other words, the first amphibians to appear on earth looked more like fish than like any amphibians that were to follow them.

A similar pattern can be observed at every stage of life's history on this planet. The earliest reptiles look remarkably amphibian-like, and the earliest mammals are actually known informally as the "reptile-like mammals." They were preceded by a group called the "mammal-like reptiles," lest the point be missed. Within other groups of organisms the pattern is even more evident, especially for animals in the more recent past, for which the fossil record is more complete.

One of the best-known examples is the horse family (scientifically, the family Equidae), whose apparent ancestry is exquisitely documented in the rich fossil sediments of North America (figure 3.1). The earliest recognizable horses appeared there nearly 55 million years ago and were relatively small animals that browsed on bushes and other vegetation. Some of them were no larger than house cats, while the biggest weighed as much as 45 pounds—about the size of an average dog. Over the next 30 million years members of the horse family ranged from 20 to 110 pounds as new species

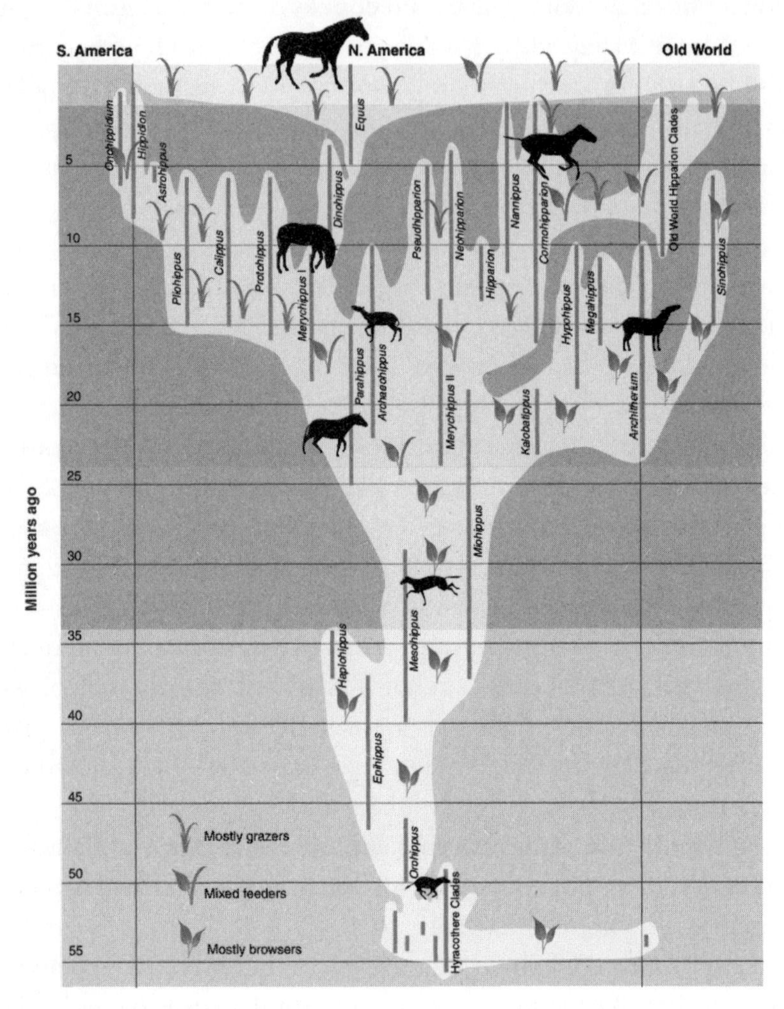

Figure 3.1: Evolution of the modern horse. A simplified representation of fossil data documenting the evolution of the horse family. More than twenty-five distinct genera are known from fossil specimens, and their likely relationships are indicated. *(Courtesy of Professor Bruce MacFadden, University of Florida. From Bruce J. MacFadden, "Adaptive Radiation of a Beloved Icon,"* Science *308 (2005): 1728–30. Reprinted with permission of AAAS.)*

emerged, and during the past 20 million years this diversification continued, leading to as many as ten species, which include the modern horse as well as zebras and donkeys.

As Bruce MacFadden of the University of Florida, perhaps the world's leading expert on fossil horses, has pointed out, the irony of this lineage is that all modern species of the horse family fit within a single genus, *Equus*. In the past, as MacFadden notes, the equine family tree actually included "some three dozen extinct genera and a few hundred extinct species."[1] What this means, of course, is that the biological diversity of the horse family was much greater in the past, when many genera (instead of just one) roamed the earth. All but one of the branches of the horse family tree have been pruned by extinction.

MacFadden leaves no doubt as to his interpretation of the equine fossil evidence. He regards it as a classic example of evolution, and of the very process for which Darwin's opponents claim there is no evidence—macroevolution. As he explains: "Macroevolution is the study of higher level (species, genera, and above) evolutionary patterns that occur on time scales ranging from thousands to millions of years. The speciation, diversification, adaptations, rates of change, trends, and extinction evidenced by fossil horses exemplify macroevolution." And indeed, they certainly seem to. But couldn't the very same patterns be explained by intelligent design?

Of course they could. Because an intelligent designer, after all, could make just about anything, he, she, or it could have designed each of those "few hundred extinct species" of the horse family and then watched nearly all of them go extinct. But taking design seriously, just like taking evolution seriously, demands that we look more closely at the patterns of species appearance in the horse family and see if they make sense in terms of design.

Evolutionary biologists ask exactly such questions. Work done by MacFadden and many others, including Christine Janis and Maeve Leakey, has carefully tracked the changes in vegetation that took place during the periods when these species appeared and became extinct. Although all the data are not yet in, many of the key events in the equine fossil record seem to be associated with climate-linked shifts in the types of grasses available for grazing in the tropical and temperate regions inhabited by these animals. A link between climate, food source, and species succession is just what might be expected to drive evolutionary change.

What, however, would be expected to drive a fossil record of this sort if it were produced by intelligent design? Incredibly no ID theorist I know of has even bothered to address the question. Considering the detail with which evolutionary biologists have studied equine evolution, the lack of curiosity from the design community about such lineages is striking. One might almost say that their primary goal seems to be to stake a claim for design, not to understand it.

Taking the principles of design seriously, however, allows us to fill this vacuum, at least in general outline, with two key points drawn from the design literature. The first point is that ID theorists generally claim that they are not young-earthers, and therefore that they accept what geologists tell us about the age and history of the planet. The second point is that, while they acknowledge that small evolutionary changes—"microevolutionary" ones—are possible in the normal course of development, major changes—of the sort they call "macroevolutionary"—are not. Macroevolutionary events, therefore, are actually instances of design. Most ID proponents would classify the formation of a new species as a macroevolutionary event, and nearly all would surely apply the term to the appearance of a new genus. Taken together, these two points mean that the designer has acted at least three dozen distinct and separate times to design new species in the last 55 million years—and that's just in the horse family.

One cannot help but wonder what these design events would have looked like to a casual observer on the scene. Depending upon the designer's methods and preferences, one could suppose that each new species might appear in an instant, maybe even in a sudden puff of smoke; alternatively, the new species might result from genetic tinkering in utero, so that a female from an existing species might give birth to a youngster representing an entirely new one. Either way, hundreds, perhaps thousands, of such events would have to take place nearly simultaneously in order to produce a new species (or genus) with sufficient genetic diversity and in great enough numbers to enable mating and reproduction. The formation of a new species by design must have been a stunning event to witness.

What does design theory tell us about the details of the horse family over the past 55 million years? First, it would not consider it a family at all. From the ID perspective, the relationships detailed in figure 3.1 aren't real, because descent with modification, which is another name for evolution, never actually took place. Those ancestor-descendant relationships so apparent to paleontologists are just an illusion. In fact, the evolutionary tree leading to modern horses isn't a tree at all, but just a collection of individual species, directly created by the designer, each without any relationship to the other.

What happened over time is that the designer created a handful of little browsing species and then, as each one went extinct, he replaced it with a modified version. When those went extinct, he drafted another round of replacements, and then another, then another. Whatever one can say of this designer, he's persistent. He's also not very skillful, since just about everything he creates goes extinct relatively soon after its first appearance. Unless, of course, constant extinction is part of his master plan.

Take the incessant nature of the designer's work in just this one small family, and multiply it by each of the scores of mammalian families. Then do it again for birds, for reptiles, and for each of the other vertebrate classes. That's only a start, of course, because we must take into account each of the animal phyla, including the innumerable families of insects. We still have plants and fungi and hosts of microorganisms to contend with. Taking design as a serious scientific theory demands that we apply it again and again to every group of organisms on the planet. What can we tell from this extraordinary project? I suggest that it leads to two inescapable conclusions regarding the work of our master designer.

The first is that he's not really a designer but a creator. We may say that a particular organ or species or biochemical system was "designed," but "design" isn't really what we mean. We may know that a building was designed, but we know that only because we can see the building itself. A design is nothing more than a concept, a plan, and we would have no evidence that a design ever existed unless someone had taken it and used it to produce a concrete object that we observe and study—unless, in other words, he had actually built the building. Similarly, in the biological realm, a bacterial flagellum wasn't just designed—it was created. By any reasonable use of language, our designer of molecular machines is actually the creator of those machines and the genes that specify them.

The second conclusion is that the designer is never satisfied—or perhaps can never get it quite right. To say that our living world is the product of design doesn't quite do it justice. Today's world was preceded, after all, by a succession of ages of biological change. If our own world is the product of design, then so were the worlds of the Cambrian, the Devonian, the Permian, the Jurassic, and every age in between. Our designer doesn't just design; he does it again and again—and his designs don't last. For all of his intelligence, the most striking aspect of his work is its impermanence. His creations are swept away time and time again by extinction, requiring him to stock the pond repeatedly to keep life going.

The inescapable conclusion that comes from honestly applying the idea of design to the fossil record is that the great intelligence behind ID is a serial creator. He brings into being new species again and again, inexplicably fashioning each one so that it bears a striking resemblance to a species just lost to extinction. In other words, intelligent design is actually a hypothesis of progressive creationism.

To be fair, many of the advocates of intelligent design try their best to argue that they are different from old-fashioned biblical creationists. If you apply ID to the geologic ages and the fossil record, it's clear that the designer didn't work over a six-day period. Taking ID seriously means acknowledging the fact that the designer worked over billions of years and was kept busy during that time creating each and every species, each biological novelty, each new gene, pathway, and biochemical machine. But he was still a creator; there's just no other way to describe it. The charge that intelligent design is just another form of creationism may be resented by many in the movement, but it's the unavoidable conclusion of taking design seriously as a scientific idea.

DECONSTRUCTING THE MOUSETRAP

The concept of irreducible complexity is at the very core of intelligent design. Irreducible complexity tells us not only that a few specific micromachines like the bacterial flagellum were designed, but that biological complexity itself is the handiwork of design. It provides us with a general principle from which we can understand each living cell as a modular system into which the designer has inserted a series of carefully matched, irreducibly complex components. Life, we might say, is like a series of mousetraps, designed from a series of interlocking parts to serve specific purposes that reveal the plan and intent of the designer.

The mousetrap, ID's favorite real-world example of an irreducibly complex machine, has been chosen to make this rather complex biochemical point comprehensible to the general public. Even though it consists of just five parts, as we've seen, all five of those components have to be in place before any mice can be caught. A mousetrap missing a piece or two is just a collection of useless parts, right?

Well, maybe not. The first time I read about the mousetrap as an irreducibly complex machine, something about the argument bothered me. I couldn't quite put my finger on it, but somewhere, I was almost certain, I had seen mousetraps used for another purpose—and a very un-mousy one. In the best tradition of scientific embrace, I began to think about the mousetrap not as an analogy but as an example of irreducible complexity itself, and wondered if the example really worked. The answer to my dilemma came when one of my daughters asked me about study hall.

I grew up in the 1950s and 1960s in a semi-industrial town in New Jersey, one of those places on the borderline between urban decay and expanding suburbia. Rahway struggled to educate its growing population of school-age baby boomers and, like many towns, had to squeeze far too many kids into aging school buildings. The results included crowded classrooms, split sessions (in which the same school served as high school in the morning and middle school in the afternoon), and the use of the cafeteria and auditorium as study spaces. I had told my daughter not to complain about the oppressiveness of her mandatory library study period by giving her one of those, "Why, when I was a kid..." lectures. I couldn't talk either, I told her, in my study hall periods, but I also wasn't allowed to stand up and browse the library books, as she was able to. "In the library? I envy you," I lied, hoping to convince her that she really did have it better than her old man had.

And then I remembered. On those days when an ineffective teacher or, worse yet, a substitute was supervising study hall, things often got out of hand. Kids would hold books in their laps where the teacher couldn't see them, and dozens would slam them shut at once, shattering the quiet of the auditorium. Then it got worse. The students sitting in the balcony would fashion paper airplanes (my specialty) and toss them into the crowd below. The teachers seldom noticed them until they had reached midflight, making it impossible to determine who had thrown them. Soon spitballs began to fly, showering down on the unsuspecting students on the floor of the auditorium.

All of us were aware of the tactical disadvantage of sitting on the auditorium floor, where one couldn't fire back at the balcony sitters without standing and revealing oneself to the supervising teacher. We were defenseless—until the mousetrap came along. One of my classmates had struck upon the brilliant idea of using an old, broken mousetrap as a spitball catapult, and it worked brilliantly. He fashioned large, floppy spitballs and carefully loaded them onto the hammer, pulled it back, and fired it over his shoulder up at the unsuspecting balcony dwellers. You should have seen the surprised looks on their faces as spitballs came zooming up into balconyland with the force of ballistic missiles.

Sadly the new weapon didn't last very long. The balcony dwellers seemed to think of this new device as an unfair alteration of the balance of power and quickly ratted on my enterprising friend, who was hauled off to the vice principal's office for this terrible violation of study hall decorum. The mousetrap spitball launcher had been too effective for its own good. And now the memory of that device stuck in my mind. It had worked perfectly as something other than a mousetrap. Perfectly.

But how could it have? Weren't the parts of irreducibly complex machines supposed to be useless until the entire machine had been assembled? If I remembered correctly, my rowdy friend had pulled a couple of parts—probably the hold-down bar and catch—off the trap to make it easier to conceal and more effective as a catapult. What was left behind was, most likely, just three parts—the base, the spring, and the hammer. Not much of a mousetrap, but a helluva spitball launcher (figure 3.2). And then, setting those happy memories of study hall aside, I realized why the mousetrap analogy had bothered me. It was wrong. The mousetrap is not irreducibly complex after all.

While it is absolutely true that my friend's three-part spitball launcher wasn't going to catch many mice, that's not the point of the argument from design. The reason that irreducibly complex biochemical machines are unevolvable is that their parts, all their bits and pieces, should have no function until they are fully assembled into the final, carefully designed machine for which they are intended. That's why natural selection cannot produce such machines—natural selection, as Michael Behe has pointed out, can only select for things that are already functioning. The same is true of the mousetrap. But if the parts of a mousetrap can have functions unrelated to catching a mouse, the mousetrap cannot be irreducibly complex.

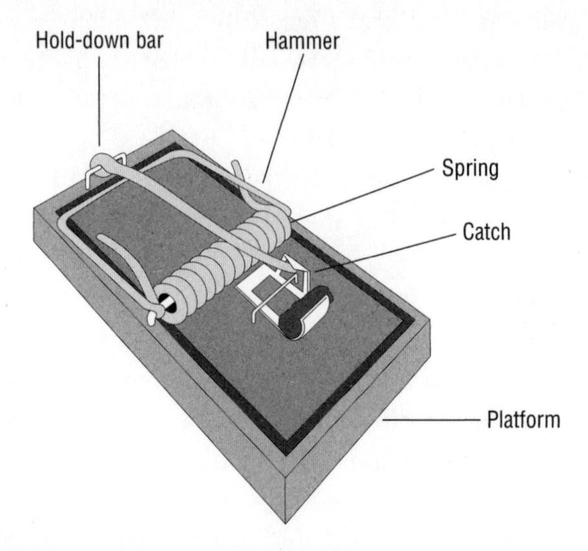

Figure 3.2: A five-part mechanical mousetrap. The mousetrap is often used by advocates of intelligent design as an example of an irreducibly complex machine. It consists of five distinct parts, all of which must be present to catch a mouse in the usual way. However, a five-part mousetrap can be easily modified to produce a three-part spitball launcher by removing the catch and the hold-down bar, demonstrating that even a partially assembled machine may be fully functional in a different context.

Give it a little thought, and you can come up with other uses for bits and pieces of a mousetrap. From time to time, just to make the point, I now use a three-part mousetrap (one just like the spitball catapult) as a tie clip. Detach the spring from the clip, and you've got a two-part machine that works as a key chain. Glue my tie clip to a sheet of wood, and you've got a clipboard. Attach a magnet, and you've got a refrigerator clip. Take the hold-down bar off, and you've got a toothpick. There are, in fact, a world of uses for little machines that include components of the mousetrap, and even more that combine mousetrap parts with other parts, such as levers and magnets.

It does, of course, take some intelligence (although not a great deal) to devise these uses and then to modify a mousetrap to make them possible. And it is also true that none of these clever little machines suggests how a mousetrap might have evolved, since, as we know, mousetraps didn't evolve; they were designed. But the most important part of ID's mousetrap argument—the contention that function is lost when any part of an "irreducibly complex" system is removed—fails. As my combination tie clip/spitball launcher shows, it's simply not true. Function may change, but in nature a changed function can still be favored by natural selection—and that's a telling point.[2]

The test, of course, will be to determine to what extent the marvelous machines of the cell resemble the mousetrap. If, like the mousetrap, their individual parts have useful, selectable functions distinct from the complete machine, then they aren't irreducibly complex. As we continue to embrace design, that's got to be the next consideration on our list. When we open up the cell and look at a biochemical mousetrap, will we find the equivalent of a spitball catapult inside it?

THE POISON PUMP

If the mousetrap is ID's favorite mechanical analogy, then the bacterial flagellum is clearly its favorite biological example of irreducible complexity. As we saw in chapter 2, the flagellum is one nifty machine—a reversible, high-torque rotary engine, powered by ion gradients and capable of propelling a bacterial cell with all the dispatch of a miniature outboard motor.[3] One could, as creationists have been doing for decades, point to its mere complexity and profess profound skepticism that it could have been produced by random mutation and natural selection. In fashioning an argument against evolution, one might well pick nearly any other cellular structure and claim—correctly—that its origin has not yet been explained in detail by evolution.

Such arguments are easy to make, of course, but the nature of scientific progress renders them far from compelling. The lack of a detailed current explanation for a structure, organ, or process does not mean that science will never come up with one. As an example, consider the question of how the left-right asymmetry that is part of our body plan arises in development, a question that was beyond explanation until the 1990s. Back then one might have argued that the body's left-right asymmetry could just as well be explained by the intervention of a designer as by an unknown molecular mechanism. Only a decade later that actual mechanism was identified,[4] and any claim one might have made for the intervention of a designer would have to have been discarded. The same point can be made, of course, with respect to any structure or mechanism whose origins are not yet understood.

The power of ID's story about the bacterial flagellum is that it seems to rise above an argument from ignorance in which we choose the explanation of "design" because we cannot imagine anything else. By asserting that it is a structure "in which the removal of an element would cause the whole system to cease functioning,"[5] the flagellum is presented as a molecular machine whose individual parts must have been specifically crafted to work as a unified assembly. The existence of such a multipart machine therefore provides genuine scientific proof of the actions of an intelligent designer.

In the case of the flagellum, irreducible complexity means that a minimum number of protein components, perhaps thirty, must be present to produce a viable biological function. These individual components should have no function until all thirty are put into place, at which point the flagellum spins into action. What this means, of course, is that evolution could not have fashioned those components a few at a time, since they do not have functions that could be favored by natural selection. As Behe wrote, "Natural selection can only choose among systems that are already working,"[6] and an irreducibly complex system does not work unless all of its parts are accounted for. The flagellum must therefore have been designed and crafted to function as a single, irreducible unit. Case closed?

Maybe not. In the popular imagination bacteria are "germs"—tiny microscopic bugs that make us sick. Microbiologists smile at that generalization, knowing that most bacteria are perfectly benign, and many are beneficial. Nonetheless, there are indeed bacteria that produce diseases, ranging from the mildly unpleasant to the truly dangerous. Pathogenic, or disease-causing, bacteria threaten the organisms they infect in a variety of ways, one of which is to produce poisons and inject them directly into the cells of the body. Once inside, these toxins break down and destroy the host cells, producing illness, tissue damage, and sometimes even death.

In order to carry out their diabolical work, these bacteria must not only synthesize powerful and deadly toxins, but must also produce an apparatus to inject them efficiently into the cells of their hosts. They do so by means of any number of specialized protein secretory systems. One, known as the type III secretory system (TTSS), enables bacteria to pump proteins directly into the cytoplasm of a host cell. The proteins transferred through the TTSS include a variety of truly dangerous molecules, among which are the "virulence factors" that are directly responsible for the pathogenic activity of some of the most deadly bacteria in existence. The TTSS is, in effect, a poison pump that bacteria use to kill other cells.

At first glance this nasty little device would seem to have little to do with the flagellum. However, molecular studies of proteins in the TTSS have revealed a surprising fact—the proteins of the TTSS are remarkably similar to the proteins in the bottom portion of the bacterial flagellum. As figure 3.3 shows, these similarities extend to a cluster of closely associated proteins found in both of these molecular "machines," and on that basis, it's now clear that the flagellum itself can be regarded as a type III secretory system.[7] In technical terms researchers have noted that the two systems "consist of homologous component proteins with common physico-chemical properties."[8] In other words, about ten of the full complement of thirty or so proteins in the flagellum function perfectly well as the TTSS. In plain language, the TTSS is just like my spitball catapult—a small part of a larger system that works just fine for an entirely different purpose.

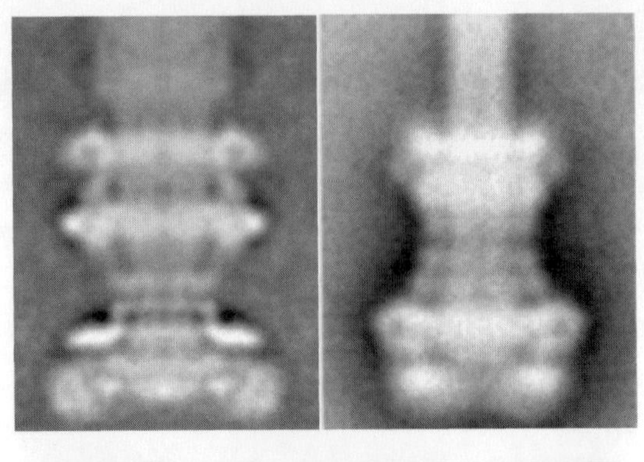

Figure 3.3: The bacterial flagellum and the type III secretory system. Electron microscope images *(top)* and computer-drawn representations *(bottom)* of the bacterial flagellum *(left top and bottom)* and the type III secretory system, or injectisome *(right top and bottom)*. The structural similarities between the base of the flagellum and the base of the injectisome are striking. The reason for this is that the proteins of the injectisome are nearly identical to proteins in the flagellar base. The injectisome, even though it does not produce rotational movement like the flagellum, acts as an effective protein-injecting system. The fact that a subset of proteins from the flagellum is fully functional in the injectisome shows that the flagellum is not, as advocates of intelligent design claim, irreducibly complex. (Top left: *Provided by Professor David DeRosier, Brandeis University.* Top right: *Provided by Professor Ariel Blocker, Oxford University.* Bottom left and right: *WGBH Television, from the* NOVA *science series.)*

Now we can see the importance of the TTSS. It means that the flagellum is not irreducibly complex.

Given that science is the search for knowledge, one would think that the design community would have welcomed all this new information about its favorite icon. That hasn't been the case. In fact, ID proponents have reacted with great fury at any suggestion that the bacterial flagellum has lost its value in making the case for design. Some have vainly argued that *both* the flagellum and the TTSS are irreducibly complex, which makes sense only if one redefines the word "irreducible." Others have pointed to speculation that the TTSS evolved from the flagellum as proving that the smaller system couldn't be the actual ancestor of the flagellum. That might be so, but other scientists have suggested quite the opposite, and in any case, it actually doesn't matter if today's TTSS is the ancestor of anything. It is absolutely correct to argue that the existence of the TTSS today doesn't answer the question of how the flagellum actually evolved. But that overlooks why the ID camp believed that irreducible complexity was so critical in the first place: They thought it proved that the flagellum couldn't have evolved. It doesn't.

It's quite one thing to point to the flagellum and challenge evolution by saying, "Evolve this!" But intelligent design has always claimed something more, something that sounds testable and scientific. It claims to have discovered a truth (irreducible complexity) that makes biochemical machines like the flagellum unevolvable, even in principle. However, once one discovers a useful, selectable function in part of such a machine, that claim falls apart, and that's exactly what has happened to the flagellum as an argument against evolution.

New research suggests, in fact, that nearly every one of the proteins in the flagellum shows significant homology to proteins that perform important functions elsewhere in the cell. In addition to the ten proteins found in the TTSS, another group of proteins in the flagellum belongs to the "axial protein" family, several are related to another secretion machine known as the type II apparatus, two are a close match for ion transport proteins, and a half dozen are involved in signal transduction pathways. In short, the flagellum isn't the custom-made, designed-from-scratch collection of closely matched elements that ID likes to claim. It's much more like a collection of borrowed, copied, and jerry-rigged parts that have been cobbled together from the spare-parts bin of the cell. In short, it's exactly the sort of thing that you'd expect from evolution.

DISSECTING RUBE

A scientific idea usually doesn't rise or fall on one example, so the demise of the flagellum doesn't mean that there might not be other examples of irreducible complexity. The blood-clotting cascade, that intricate pathway of enzymes and cofactors, looks like a good place on which to fall back, so let's go there next. Every protein in the pathway is essential to the clotting process. As Michael Behe has written, the entire system has to be in place for clotting to work properly, and in the absence of *any* of the components, blood does not clot and the system fails. That should make it impossible even to imagine how the system could have arisen by evolution. To reinforce the case for design, we might draw upon the enormous amount of whole-genome analysis that has been done using high-efficiency DNA sequences techniques. What we should find, as ID theorists have made clear, is that the complete vertebrate blood-clotting pathway is present in the genome of each and every organism we analyze.

It's a bit of a surprise, therefore, to discover that ID advocates have not been vigorously combing the databases of vertebrate genome sequences to reinforce their argument. After all, what better way to establish the irreducible complexity of this system as a genuine scientific principle than to show that their bold predictions about the need for *all* of the components to be present are absolutely true? Unless, of course, those predictions are wrong—which now clearly seems to be the case.

An old report from the 1960s did suggest that whales and dolphins lacked one of the clotting factors,[9] but ID advocates could easily have explained that away as an unreliable product of research in the premolecular age of genome analysis. In 2003, however, there was no possibility of that defense when an even more sweeping discovery was made: The genome of the fugu, or puffer fish, lacks three of the clotting factors—and its blood clots just fine.

In many respects the discovery of a functional clotting cascade lacking three components does even greater damage to the concept of irreducible complexity than the type III secretory system did to the flagellum. The existence of a partial pathway that not only has a useful function, but performs what we might call the final function (blood clotting) demonstrates beyond any doubt that complex pathways can be built up a few steps at a time from simpler ones. It also dashes any hope that the components of the clotting pathway might be seen as a Rube Goldberg collection of custom-made parts. In fact, it's something quite different, as further studies of several animal genomes have now shown.

We could nonetheless still challenge evolution by asking the question of whether it could produce such intricate pathways, even ones that are not irreducibly complex. Here's where modern protein chemistry, genetics, and genome analysis provide an interesting answer.

One of the striking things about many of the clotting proteins is how similar they are. The key ones belong to a group of enzymes known as the serine proteases, and for several decades researchers have speculated that they might have been formed by a process known as gene duplication. Every now and then, when DNA molecules are duplicated just before a cell divides, the machinery makes a mistake and copies the same small section twice. Although having a duplicate copy of certain genes can get an organism in trouble, most gene duplications don't cause problems. Rather, they produce opportunities, some of which may have given rise to the clotting factors.

A detailed analysis of the clotting factor proteins themselves gives credence to this suggestion. The long chains of amino acid building blocks from which proteins are formed fold into distinct patterns that protein chemists call "domains." A domain is something like one of the unique Lego blocks that are used for special purposes, such as holding the propeller on a model airplane or forming the hand on a toy soldier. By swapping those parts back and forth, one can build new toys and models, but the specialized parts always remain recognizable because they are so different from the ordinary blocks that make up most of the kit. Protein domains are similarly specialized and recognizable and, just like Lego blocks, can be swapped back and forth, in their case to make different proteins. As luck would have it, only a small handful of domains are used to build most of the clotting factors. One of those, naturally enough, is the serine protease domain.

Decades of work analyzing these proteins have led to the recognition that just six gene-swapping events could have produced nine of the clotting factors shown in figure 2.1. This conclusion comes from work done in the laboratory of Russell Doolittle, supported by many other research labs that have followed his lead in studying the clotting proteins. The steps in the pathway, which require swapping domains in and out of the genes that specify the proteins, involve nothing more than ordinary genetic changes of the sort routinely observed in living organisms. Therefore, there is no reason to invoke the hand of a designer-creator or any other force outside of nature to account for the pathway.

Can we prove that this is how the clotting pathway evolved? Definitely not. But that's hardly surprising, because experimental science simply isn't in the business of "proving" things. However, we can and do test ideas such as these, and the rapidly increasing number of whole-genome DNA sequences makes rigorous tests possible. If the clotting factors did indeed evolve this way, then the detailed sequences in each of them should show a pattern of relatedness that exactly fits the scheme. Putting it another way, if two of the proteins, such as factors 9 and 10, diverged millions of years ago, as the pathway suggests, then we should be able to take any organism we like and discover that the pattern of relationships between their various factor 9 proteins matches the one we see for their factor 10 proteins, which similarly matches the one we see for factor 7. A designer, of course, would have designed the proteins to clot blood, not to confirm a set of putative evolutionary relationships. ID, therefore, cannot predict such patterns, and cannot even be tested, since arbitrary design (we cannot understand the designer's plans) could be consistent with anything. So, what do we actually see? Analysis of the detailed sequences of these proteins matches the prediction of evolution perfectly—it passes the test.

Now let's ask another, more fundamental question. Even if we've been able to show that the pathway is not irreducibly complex, since so many organisms have been found in which clotting occurs in the absence of several of the factors, where did the original factors come from? Surely a system of several genes could not have simply appeared out of thin air, unless, of course, the system was created by an intelligent designer.

As we've seen, the clotting factors are made up of modules called protein domains. Evolutionary theory would suggest that an ancestral vertebrate lacking today's clotting pathway would still have had most of those modular protein domains, perhaps performing other functions, scattered around its genome. But how could we check that possibility, since that ancestor lived more than 400 million years ago? The answer to this question provides a second stringent test for the evolutionary explanation. If these protein domains were present prior to the evolution of today's clotting pathways, then living organisms descended from that nonclotting ancestor should still contain some of those domains. The sea squirt *(Ciona intestinalis)* is exactly such an organism.

Sea squirts don't have backbones, so they are not vertebrates. However, in their larval stages they do have a notochord, a long supporting rod that runs along the back of the body, just below the nerve cord. This makes them chordates, in the jargon of biology. Humans are chordates as well, along with all other vertebrates, since we also possess a notochord during our embryonic development. Today's sea squirts are not our ancestors but are descendants of organisms that split off from the line leading to us and to all other vertebrates, including (of course) those with the blood-clotting pathways. So, if the modular domains of our clotting system were formed by mixing and matching preexisting domains from other genes, then there's a pretty good chance that the genome of the sea squirt ought to still have a few of them left. Does it?

In 2002 the complete genome of the sea squirt was determined,[10] and it became possible for the first time to look for those domains. The results were stunning. Even though none of the sixteen thousand genes of the sea squirt codes for anything resembling a vertebrate clotting factor, copies of all but two of the protein domains from which those factors are built were found scattered around its genome. In effect the sea squirt has nearly all of the nuts and bolts, all of the spare parts, that would have been necessary to piece the first clotting factors together 400 million years ago. A designer, of course, could have created the clotting pathway from scratch, which is exactly what the proponents of intelligent design claim. But if that were indeed the case, then why do we find the raw materials for clotting exactly where evolution tells us they should be, in the last group of organisms to split off from the vertebrates before blood clotting appeared? Once again, evolution passes the test, and advocates of design are left to explain why the designer would have wanted to produce a pathway that looks as though it evolved.

SAME OLD, SAME OLD

More recently, Michael Behe has refocused his argument for intelligent design by attempting to use statistics against evolution. His 2007 book, *The Edge of Evolution,* examines the subject at the level of genes and proteins. His case study is the continuing life-and-death struggle between our own species and *Plasmodium,* the parasite that causes malaria. Behe concedes, as he must, that evolution has changed both parasite and host. Drugs that once controlled the disease, such as chloroquine, are now ineffective because of evolution within the parasite population. Evolution has also produced new forms of resistance to *Plasmodium* within the human population, just as any biologist would predict. In fact, so well does Darwinian evolution account for these changes that, as Behe puts it, we should score one for the Sage of Down House.

But, as he argues, that's about all we can depend upon evolution to accomplish. In Behe's view, these are examples of nothing more than a kind of "trench warfare" in which the two species have progressively disabled or broken parts of themselves in order to survive. Nothing genuinely new, novel, or complex has resulted from this struggle, and we shouldn't expect otherwise. The reason, according to Behe, is that the sorts of changes we see in this well-studied interaction represent the limit, the "edge" of what evolution can accomplish. They can go this far and no further. A line in the sand is drawn, and on the other side of that line is intelligent design.

How does Behe know where to draw that line? He takes a rough estimate from a 2004 clinical paper[11] as to how often resistance to the antimalarial drug chloroquine has arisen in natural populations. Noting that several mutations are necessary for resistance to develop, the author of that study, Nicholas J. White, suggested that the odds of their occurring in a single parasite would be 1 chance in 10^{20}. Why does this matter? To Behe, the collection of mutations required to confer maximum resistance to chloroquine stands as a mark of complexity in a protein. In fact, he even invents a term for it, calling it a "chloroquine complexity cluster," or CCC. He suggests that the odds of any protein developing a similarly complex feature, such as a binding site for another protein, would be on the order of 10^{-20}. To convert this number into an argument against evolution, Behe engages in a sleight of hand reminiscent of his argument for irreducible complexity.

Cells are filled with complex protein-to-protein interactions in which one protein binds, or attaches, to another. Signaling pathways work this way, with one protein binding to another and that protein to another and so forth. Molecular motors and transport machinery within the cell also depend upon protein-to-protein binding sites. If the average protein binding site has the complexity of a CCC, then the odds of evolution producing just one new binding site are, according to Behe, 10^{-20}. How about two binding sites, the minimal requirement for even a simple signaling pathway? Behe squares the odds, making the chances of a new two-protein interaction produced by evolution equal to the product of $10^{-20} \times 10^{-20}$, or 10^{-40}. Given Behe's estimate that fewer than 10^{40} cells have existed during the entire history of life on earth, that means that the evolution of even a moderately complex system of interacting proteins is far beyond the "edge" of what evolution can accomplish. He is so certain of this conclusion that he calls it the two binding sites rule. Wherever we see two binding sites in a protein, he assures us, we see the hand of design.

Hidden within these numbers is the same flawed argument that doomed irreducible complexity. Behe's math requires that all of the mutations that produce a CCC must occur together, and they can be favored by natural selection only when all of these highly improbable events take place. This formula, however, ignores the realities of chloroquine resistance in malaria. Molecular studies of drug resistance in the parasite[12] show that it is not the all-or-nothing, one-chance-in-10^{20} event he claims. Rather, full-blown resistance is preceded by a number of mutations that confer *partial* resistance, enabling natural selection to work at every step of the process.[13] This means, as Nicholas Matzke wrote in his review of *The Edge of Evolution*, that chloroquine resistance "is both more complex and vastly more probable than Behe thinks."[14]

The same criticism applies to the way in which Behe squares these odds to obtain his two binding sites rule. In calculating probabilities, we multiply individual odds of two events only if they are completely independent of each other, like the simultaneous flipping of two coins. By squaring 10^{-20}, Behe assumes that both binding sites must appear at the same time in a single individual cell by pure, random chance. What he ignores, of course, is something that we already know to be true in the case of a CCC, namely that natural selection can favor intermediate stages on the way to the evolution of the final, fully resistant organism. Behe's two binding sites rule makes two critical assumptions: (1) that all of the features of each binding site must appear simultaneously, against great odds;

and (2) that each binding site is absolutely useless until the second appears. In fact, both of these assumptions are wrong.

Significantly, Behe's "new" arguments are remarkably similar to his old ones. He stacks the odds against evolution by contending, against all evidence, that only the final form of a structure (or a pair of binding sites) has any selective value, and then pronounces it impossible to achieve by evolution. Behe is not willing to consider the possibility that a weak binding site, produced by just a few changes in an existing protein, might appear first and then evolve through natural selection into a more specific interaction. The error in logic is identical to the earlier claim of irreducible complexity. One demands that every part of a biochemical machine, or a protein binding site, must be assembled simultaneously for the machine to work or the site to bind. Evolution then seems to be incredibly improbable.

In reality, study after study has shown that these demands are not realistic. Researchers reporting on their detailed study of the evolution of critical binding sites in two different receptor proteins put it this way:

> The puzzle that complex systems pose for Darwinian evolution depends on the premise that each part has no function—and therefore cannot be selected for—until the entire system is present. . . . But virtually all molecules can and do participate in more than one process or interaction, so a complex's elements may have been selected in the past for unrelated functions. Our work indicates that tightly integrated systems can be assembled by combining old molecules with different ancestral roles together with new ones.[15]

In other words, even the "new" argument for design depends upon the same false premise. Little wonder that researchers have brushed ID aside and are busy instead exploring the world of molecular evolution.

JUST NOT GOOD ENOUGH

We could go on. There are a host of complex systems that ID advocates attribute to the work of the designer, and given paper, time, and patience—more patience than you or my editor might be willing to show—we could explore each and every one of them. What we would learn, however, is exactly what we have seen thus far in the cases of the bacterial flagellum and the blood-clotting pathway. The arguments advanced by advocates of ID as to why these structures or pathways could not have evolved turn out to be wrong, and in many cases the processes by which they were formed can be tested and examined.

Why, then, does the intelligent-design movement continue to present these systems as examples of biological micromachines that provide scientific evidence for special creation by a designer? They might, of course, rightly point out that until the step-by-step evolution of each system is explained in detail, they can continue to invoke design as an explanation for them. But this would be a profound retreat to the old arguments of the eighteenth and nineteenth centuries, which attributed any scientific mystery, any unexplained natural process, to the direct involvement of the Almighty. The great claim of the ID movement, however, is to have risen above mere personal incredulity, reaching a point where design could be scientifically detected and analyzed.

If there was ever a moment in the past few years when the ID movement was given the chance to display that level of scientific sophistication, it came during a federal trial in 2005. The school board of the small town of Dover, Pennsylvania, had decided that intelligent design was a credible and genuine scientific theory, and proceeded to tell students about it and to provide them with reading material on ID. That decision landed them in court, opposed by a group of parents who not only disagreed, but saw the board's policy as religiously based. The board's attorneys argued that instructing students about ID served the legitimate secular purpose of promoting science education and critical thinking. The judge in the case gave both sides wide latitude in presenting evidence on this point, and the defense team representing the board brought leading scientific spokesmen for ID, including Dr. Michael Behe, into the courtroom as expert witnesses. Several years earlier Behe had written that a key portion of the immune system would never be explained by evolution. He used this, of course, as yet another example of an irreducibly complex system that could not have evolved and therefore provided powerful evidence for design.

A few words of explanation are in order. Our immune systems fight disease in many ways, but one of the most important is by producing proteins known as antibodies. A key portion of each antibody, known as its combining site, attaches to molecules on the surfaces of bacteria, viruses, or foreign cells. Once a few antibodies stick to one of these pathogens, it has been effectively marked for the rest of the immune system to destroy it. The tricky part, as you might expect, lies in the way in which the immune system builds tens of thousands of different antibody molecules, each with a slightly different combining site, so that one will attach to a smallpox virus, another one to a polio virus, and another one to just about any type of molecule found on an invader.

For decades how the immune system produced so many different types of antibody molecules remained a mystery, but in the 1970s it was solved in a brilliant series of investigations by Susumu Tonegawa of MIT.[16] His work showed that certain cells in the immune system carry out a process of genetic shuffling in which they mix and match the very pieces of DNA that code for the combining sites of particular antibodies. The process is remarkably like shuffling a deck of cards. The hand is dealt out the same way in every cell, but the shuffling mixes up the cards so that the outcome varies at every table in the casino. Since the shuffling process produces a different result in each cell, hundreds of thousands of unique antibody genes are produced, each coding for a different antibody protein. This allows the trillions of cells in the immune system to build up a genetic library in which there will always be a few antibody-producing cells ready to produce just the right antibody protein for almost any pathogen. It's an amazing system. But how did it evolve?

In essence the system has three basic parts. In addition to the right DNA sequences, which contain the blueprints for the antibody molecule itself, the cell also needs a molecular machine that cuts and recombines the pieces of DNA during the shuffling process. And finally it needs to have a set of signals built into the DNA telling that machine where to cut and paste during the shuffling process. This multipart system is another example of the type of process that ID advocates claim could not have evolved, and therefore must be further evidence for design, because the individual parts are useless without one another. Michael Behe pointed out this problem for evolution in 1996:

> In the absence of the machine, the parts never get cut out and joined. In the absence of the signals, it's like expecting a machine that's randomly cutting paper to make a paper doll. And, of course, in the absence of the message for the antibody itself, the other components would be pointless.[17]

Behe, of course, was aware of an interesting suggestion made several years earlier by Nobel laureate David Baltimore as to how this system may have come about. Like other scientists, Baltimore noticed that the gene-shuffling system in the immune system has a striking resemblance to a class of DNA molecules known as transposons, or transposable genetic elements. Behe, however, ridiculed this suggestion, comparing it to a fanciful ride in a magic box by cartoon characters Calvin and Hobbes. In fact, he told researchers that there was simply no point in doing research into the evolution of this system:

> As scientists, we yearn to understand how this magnificent mechanism came to be, but the complexity of the system dooms all Darwinian explanations to frustration. Sisyphus himself would pity us.[18]

Fortunately scientists in the field paid no attention to this advice and continued to investigate Baltimore's idea. In 1996 they noticed strong biochemical similarities between antibody gene shuffling and the ways in which transposons move from point to point in the genome. Labs around the world followed this lead, and one step at a time they confirmed each and every element of the transposon hypothesis. These studies reached their climax in 2005, when the exact transposon from which the immune system machinery evolved was conclusively identified.[19]

How did the ID community respond to this work? Michael Behe was systematically presented with the results of the studies while on the stand at the Dover trial. Judge John E. Jones III first noted the amount of evidence arrayed before him:

> Between 1996 and 2002, various studies confirmed each element of the evolutionary hypothesis explaining the origin of the immune system. [2:31 (Miller)] In fact, on cross-examination, Professor Behe was questioned concerning his 1996 claim that science would never find an evolutionary explanation for the immune system. He was presented with fifty-eight peer-reviewed publications, nine books, and several immunology textbook chapters about the evolution of the immune system.[20]

He then remarked on Behe's reaction to the pile of research papers and books contradicting his assertion of unevolvability:

> However, he simply insisted that this was still not sufficient evidence of evolution, and that it was not "good enough." [23:19 (Behe)][21]

The judge understood that merely shaking your head to reject a substantial body of research places a "scientifically unreasonable" burden of proof on those who have elucidated the evolution of this key part of the immune system. The lesson from this part of the

Dover trial was clear to everyone in the courtroom: Even when presented with every opportunity to make their case, the defenders of design resorted to little more than saying "It's not good enough for me" in the face of overwhelming evidence for evolution.

WRITING THE BOOK OF LIFE

One of ID's most attractive elements is the way in which its proponents have couched it in the language of modern science. The arguments for ID, as they are presented, seem to be based not on a biblical or theological perspective, but, rather, on hard, cold, objective science. They cite macromolecular complexes and biochemical pathways and claim to use the most advanced scientific methodologies to detect design. Nowhere is this spirit of avant-garde inquiry more apparent than in William Dembski's law of conservation of information. Although most of us have a sense of what information is, because we deal with it every day of our lives, arguments based on information theory are beyond the training of most individuals, even those in scientific and technical professions. As a result pronouncements about the sources of biological information have a compelling ring to them. One example is the conclusion of a 2002 article by Dembski in *Natural History* magazine, in which he assures readers that his work has demonstrated that

> undirected natural processes like the Darwinian mechanism are incapable of generating the specified complexity that exists in biological organisms. It follows that chance and necessity are insufficient for the natural sciences and that the natural sciences need to leave room for design.[22]

Dembski's definition of "specified complexity" is highly technical and is explicated over the course of hundreds of pages in the several books he has written on the subject. What it boils down to is a particular definition of information that can be applied to biological systems. The information in a living cell certainly exhibits complexity, as any biologist would agree, but *specified* complexity involves something more. An event exhibits specified complexity, according to Dembski, if

> it is contingent and therefore not necessary; if it is complex and therefore not easily repeatable by chance; and if it is specified in the sense of exhibiting an independently given pattern.[23]

If specified complexity can arise only from an intelligent cause, and if living cells are full of the stuff, then surely living cells were produced by an intelligent agent. That's a powerful argument, and it certainly would revolutionize biology if it were true. In fact, it would be the discovery of the decade, if not the century. The plot of Carl Sagan's science fiction novel *Contact* was built around just such a discovery. In the novel, radio astronomers pick up a faint signal that at first seems to be nothing but noise, but quickly reveals itself to carry an unmistakable signal of intelligent origin. Coded in part of the signal are the first 261 prime numbers, and the scientists conclude that the signal is a message sent from an intelligent source.[24]

I laughed when I first read *Contact* because back in the 1970s I had toyed with the idea of writing a story along similar lines. In my imagination—the imagination of a biologist—those prime numbers would instead have been encoded in part of the human genome, but they would likewise have stood out as the clear mark of an intelligent agent. Fortunately reality set in and I turned back to my day job in the cell biology laboratory, thereby sparing myself the pile of rejection slips that would have been the result of my half-baked efforts as a storyteller.

Nonetheless, the notion that there might be a clear and distinct informational signal in human DNA is intriguing, and if we were indeed to find prime numbers, telephone directories, or the scores of the first forty Super Bowls encoded in part of our DNA, we'd certainly sit up and take notice. Alas, there are no such obvious signals. But what of the claim that the coded sequences of genes themselves constitute a kind of signal, a form of information that points directly to an intelligent agent?

If Dembski's claim that the mechanism of evolution cannot produce new information was correct, we'd be able to verify it by using computer programs that mimic evolution. There are quite a few of these around, but one of the most straightforward has been written by Thomas Schneider of the National Institutes of Health (figure 3.4). He calls his program ev, for evolution, and it starts with a set of completely random sequences of 256 bases, sequences written in the four-letter language of the DNA bases (A, T, G, and C). At each generation, it evaluates how well a protein binds to (recognizes) a collection of "binding sites," places where a small protein encoded by the same DNA sequence might bind. At the beginning of a test run, of course, they don't bind very well. Nonetheless some of these random sequences do bind a little better than others, so Schneider's program allows half of the sequences, the half that bind best, to form the next generation.[25]

That succeeding generation is formed by introducing unpredictable errors (mutations) into the successful sequences, and then testing their binding ability again. Schneider's program automatically calculates the information content of the sequence and shows how it changes as this process of selection, reproduction, mutation, and selection continues. The results are striking.

The information content of the sequence rises steadily throughout the selection process and gradually levels off at a point where the fit between protein and binding sequence is nearly perfect. The strength of Schneider's simulation is that the actual content of information can be measured mathematically in terms of bits of information. His measurement is taken according to methods developed by Claude Shannon, the father of modern information theory, and unequivocally shows that this process leads to an objective and quantifiable gain in information. To find out how this happens, Schneider allows the simulation to proceed as before, but removes the selective step in which only the best-binding sequences are allowed into the next generation. The information content of the sequences quickly drops back to zero, showing that high information content is directly dependent upon a continuing process of selection.

What's needed to drive this increase? Just three things: selection, replication, and mutation. The best sequences are selected for replication into the next generation, then they are mutated, and then they are selected again. It's no coincidence that the same three things are required for evolution, since what we are observing is nothing less than evolution on a small, observable scale.

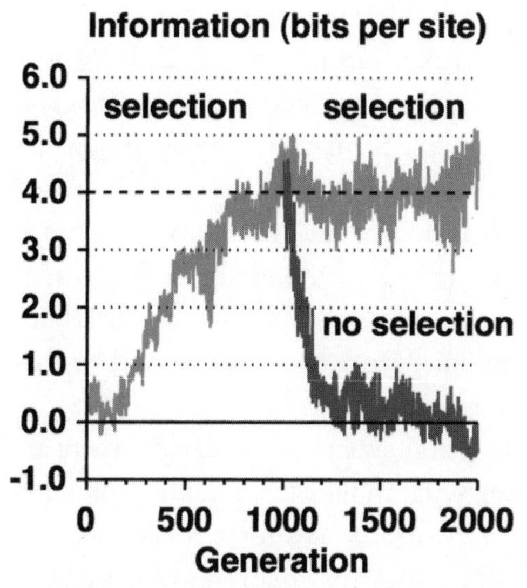

Figure 3.4: Natural selection is responsible for an increase in informational complexity. A computer simulation of evolution by natural selection shows that randomized information increases as the result of selection applied to an evolving digital "organism." The ev program, written by Thomas Schneider of the National Institutes of Health, simulates a protein binding site and demonstrates how the pressure of natural selection increases the information content of the site. Note how the information content drops when selection is stopped. *(Dr. Thomas Schneider, National Cancer Institute, National Institutes of Health.)*

Where's the new information coming from? Perhaps the investigator is sneaking it into the system at the start? No chance of that, since the starting sequences are completely randomized. Maybe there's hidden information in the program itself? Not likely. Schneider has made the source code of his program open for inspection, and there isn't even a hint of such nonsense. Did Schneider rig the parameters of the program to get the result he wanted? Not at all. In fact, changing the program in just about any way still results in an increase in measurable information, so long as we keep those three elements—selection, replication, and mutation—intact. Where the information "comes from" is, in fact, from the selective process itself.

Does this mean that evolution gives us a "free lunch," that we get something for nothing in terms of information content? No. In fact, when you think about it, a very high price is paid to produce that information and then to keep it. That price is the cost of replication and selection. Schneider's model requires an extravagant degree of waste, since half of the "organisms" (his DNA sequences) must be thrown away in each generation, and considerable energy supplied to replicate the surviving organisms, which then undergo a round of mutations to generate further variation. The information in the system is generated and preserved by this costly process of selection and replication.

In the living world evolution occurs at a comparably great cost. Well before Darwin, thoughtful people wondered at the grand extravagance of nature. They puzzled at the cost of reproduction, at the great numbers of organisms that are born or hatched or sprouted each year, especially when only a few successfully reproduce into the next generation. That great expenditure—the energy and food and effort required for survival and success in the selective process—is the ultimate source of the information we can detect in living systems. The book of life has been written, and continues to be written, by cycle after cycle of costly growth, reproduction, selection, and survival.

INFORMATION IN THE REAL WORLD

Thomas Schneider's ev program is just one of several computer-based efforts to model the process of evolution. Others include Avida, a program developed at Michigan State by Charles Ofria and Richard Lenski that follows the evolution of self-replicating computer codes, and Tierra, a program written by Thomas Ray at the University of Delaware that also models digital organisms. In every case, as the starting conditions evolve in the digital environment, the programs' information content increases, directly violating William Dembski's "law" of conservation of information. What gives?

One possible response is that all of these simulations are just computer programs, creations of the minds of those who conceived and programmed them. Isn't this itself a form of intelligent intervention? Well, not really, especially when the starting information content of the system is zero (as is the case for the ev program) and all the program does is to allow for replication, selection, and mutation, which is exactly what evolution does. But it can certainly be argued that any simulation, no matter how carefully constructed, is just a model of a living thing. The true test lies in the real world, in living organisms themselves. If evolution actually can produce new information, that's where we should see it.

Quite unintentionally the Western chemical industry has for much of the last century been carrying out an exceptional experiment on the ability of evolution to produce new information. It has done so by producing synthetic compounds, molecular forms that have never existed in the natural world, and then turning them loose. If living organisms could not produce new information by evolutionary processes to break down these compounds, they would persist in the ecosystem. But if evolution can create new genes and new proteins capable of dealing with these compounds, we would have real-world proof of the ability of the mechanisms of evolution to generate that information.

In 1935 DuPont chemist Wallace Carothers carried out a chemical reaction that joined together two chemical compounds, a diamine and a dicarboxylic acid, to produce a synthetic molecule never before seen in nature. The new compound was a polymer, a long chain in which these repeating units could be joined together again and again, to make chains of almost limitless size. The polymer was tough, flexible, and durable, and DuPont decided to call it nylon. Factories around the world began producing the new compound, and before long nylon compounds were found everywhere in the environment.

Not surprisingly, nylon, once synthesized, tended to hang around. Engineer a stable and completely new compound, as Carothers did, and you're unlikely to find an organism that can break it down. In the 1970s, however, a number of Japanese scientists made a most unusual observation. Industrial plants engaged in nylon manufacture produce chemical waste products from the synthetic reaction and often discharge these wastes into holding pools around the factories. Containing, as they did, only water and nylon waste compounds, nothing should have grown in these ponds. But something *was* growing, in fact: scummy mats of bacteria, apparently thriving on the artificial, synthetic polymer. To their very great surprise, the researchers discovered that these bacteria possessed an enzyme capable of breaking down, or hydrolyzing, nylon. They had evolved "nylonase."

Enzymes, as design advocates like to emphasize, are highly specific. They have to bind the chemical compounds they process, and then they have to carry out a chemical reaction perfectly—in this case a tricky reaction that breaks the key bond that holds the parts of the nylon compound together. Incredibly these bacteria could do precisely that, and they could do it well enough to break nylon apart into its components and then eat the leftovers as food. In effect they'd found a genuine free lunch outside the chemical plants.

Where did this new enzyme come from? Studies in the California laboratory of Susumu Ohno found the answer in 1984. The nylonase enzyme in these bacteria was derived by duplicating a gene that controlled another, quite different function, and then inserting a single extra base into its DNA code. Why would this produce so different a protein? It turns out that the genetic code is read in a very specific way, three letters at a time. You might think of it as a message like this:

THECATATETHEFATRAT

At first that string of letters doesn't seem to make much sense, but separate the letters into groups of three, and you get a sentence that any second-grader would be able to read:

THE-CAT-ATE-THE-FAT-RAT

Consider what happens, however, if we insert an "extra" letter into the message, and then try to read it, three letters at a time. If we insert a T, for example, into the fourth position of the message, we'd get

THE-**T**CA-TAT-ETH-EFA-TRA-T

Suddenly, everything that follows the inserted letter becomes meaningless gobbledygook. This, in fact, is usually what happens when an extra base is inserted into a gene. Shifting the "reading frame," as biologists call it, most often results in an absolute mess, in which one of the three-letter coded instructions for "stop" crops up and thus prevents the machinery of the cell from producing a complete protein. In this case it turns out that shifting the reading frame of the preexisting gene produces a protein nearly as long as the original with an entirely different set of properties—the "words" of the sequence had indeed been scrambled, just as in our example, but in the bacterium the sentence still had a valid, but new meaning: It produced an enzyme capable of binding and breaking the bonds of nylon. The bacteria possessing the new gene could eat nylon, and they thrived in the wastewater ponds on this previously unused source of food.

There's absolutely no doubt that any gene that could be frame-shifted in this way and still produce a functional protein would be most unusual. One could chalk this up to sheer dumb luck, and maybe even to the "design" of the original gene, which contained a series of internal repeats, making it unlikely that the accidental insertion or deletion of a DNA base would result in a "stop" instruction. But we cannot get around the fact that the information needed to break down nylon came from the very process of mutation, selection, and reproduction that characterizes evolution. In reality, generations of mutation and selection on the whole bacterial genome provided the raw material from which this one sequence has emerged to give a few individuals a selective advantage in a nylon-rich environment. And, in truth, the nylonase activity displayed by this new enzyme at this stage of development isn't very impressive, much like the weak interactions that appear in the early stages of Thomas Schneider's ev program. As is the case with the simulation, however, we should expect that natural selection will over the course of time fine-tune the enzyme to make it more and more efficient in its interactions with nylon compounds.

BEFORE OUR VERY EYES

No one ever sees design taking place. That's a point that ID advocates are fond of emphasizing whenever evolutionists challenge them to show evidence for their ideas. We cannot know the identity of the designer or the means by which the designer works. But no matter—we can't see evolution working, either. So evolution and intelligent design are both just matters of faith, worldview, or philosophy, right?

That might be the case if evolution could indeed not be observed, if it was a matter of pure inference, no matter how well supported by the evidence. But the process *can* in fact be observed directly, and under controlled conditions. The evolution of nylonase is precisely such an example. Intrigued by the unavoidable conclusion that this new enzyme had evolved naturally by means of gene duplication and mutation, about ten years ago a group of researchers at Osaka University decided to see if this remarkable process could be reproduced in the lab.[26] They took a culture of *Pseudomonas* bacteria that had no ability at all to metabolize nylon compounds and grew it on a medium containing small fragments of the nylon molecules[27] as the sole source of food.

After just nine days they found colonies of "hypergrowing" bacteria that had begun to master the trick of using nylon fragments as food. They then transferred these bacteria to a medium containing another nylon compound, and within three months they found that some of the bacteria had evolved the ability to grow on this compound as well. As the researchers summarized their results: "In the present study, it was shown that microorganisms can acquire an entirely new ability to metabolize xenobiotic compounds such as a by-product of nylon manufacture through the process of adaptation." In short, they evolve the ability right before our very eyes.

Prior to this experiment one might have asked if we could really be certain that evolution had produced this capability. Maybe those nylon-digesting genes had already been present and were just waiting for Wallace Carothers to cook up his new compound. That turns out not to have been a possible explanation for the *Pseudomonas* experiments, in which the inability of the bacteria to metabolize nylon was carefully documented before the study began. As it went on, though, week after week, evolution ran its course, and the new capability developed under direct human observation.

However we look at it, the information for this entirely new capability came about without a designer, without a plan, and from the sheer opportunism of living organisms generating new variations in the face of selective pressure. And the source of the information, in a way, is nylon itself. By introducing nylon into the environment, we provided trillions of bacteria with a puzzle for which the solution promised a great reward. Come up with an enzyme that lets you break down nylon, even if it's just a quick and dirty modification of an existing one that doesn't really work very well, and you'll be given a free lunch. Under such circumstances the information needed for the solution is found in the selective process itself, just as Schneider's program tells us.

Evolution works in the real world, even when we're watching!

EVOLVING DESIGN

The apparent design of complex molecules such as enzymes makes for a powerful emotional argument against evolution. But arguments based on first appearances are always suspect, especially in science. After all, the sun *appears* to revolve around the earth, when in reality the opposite is true. The reality of biological information is that the mechanism of evolution produces new information in response to changing conditions, and it does so in precisely the way that Charles Darwin predicted—by "descent with modification." New and novel genes emerge from modified copies of old ones, and they do so with surprising speed.

If the rapid emergence of nylonase were the only well-studied example of such rapid and profound emergence of new information, we might justifiably dismiss it as a special case. But it's not a unique example at all. As the tools of molecular biology have become more powerful, studies documenting the evolution of new proteins and even new biochemical pathways have become increasingly common.

Shelley Copley[28] described a novel pathway that breaks down a pesticide known as pentachlorophenol (PCP). That pathway is remarkable because PCP, just like nylon, is a synthetic compound first produced in the 1930s. The enzymes in the pathway were "recruited," according to her study, from two quite different chemical pathways and then modified by mutation to fit their new tasks. A similar case was reported by a group of researchers at a United States Air Force laboratory in 2002.[29] Investigating the source of a bacterial pathway that breaks down 2,4-dinitrotoluene, one of the components of explosives used by the Air Force, the researchers came across clear and unmistakable evidence that the enzymes of the pathway had evolved only recently. The pathway is indeed complex, consisting of a total of seven enzymes, all modified versions of enzymes originally used for other purposes.

One of the most dramatic examples of evolution using old genes for new purposes is that of Antarctic fish. The southern polar ocean began to freeze over a little more than 10 million years ago, producing a strong selective pressure in favor of fish that might be able to keep their blood from freezing in the increasingly chilly waters. As a result a variety of fish underwent evolutionary adaptations that modified preexisting genes to produce proteins that served as a kind of biological antifreeze. One of these has been studied in such detail that we now know exactly where it came from.[30] Evolution didn't merely take an existing gene and cobble a few changes into it. Rather, it started with an extra copy of a gene for a digestive enzyme. Mutations deleted most of the original gene, and then duplicated, over and over again (forty-one times, to be exact), a small portion at the beginning of the gene. The modified gene was then expressed in the liver, and the protein it codes for is targeted to the bloodstream. The duplicated region works as a powerful antifreeze agent, and the fish thrive in the chilly waters. The new gene is so different from the original that it might be difficult to recognize were it not for the fact that both ends of the antifreeze gene match the "old" gene of the digestive enzyme, betraying its origins and marking it clearly as the product of rapid evolution.

The list could go on, but the point is clear: Theoretical studies that model evolution show how information for new genes is produced by the process of selection, replication, and mutation. Studies of existing genes confirm that these processes work in the real world, and they provide one example after another of how evolution solves the "problem" of information by using the demands of the environment to generate it. It does so at great cost, to be sure, but the generation of new information can be observed, documented, and even studied in real time. The "design" of new genes does occur, but the designer is the process of evolution itself.

DESIGN'S SWEET TOUCH

I began this chapter by suggesting that the proposition of intelligent design should be taken seriously, and that's exactly what I have tried to do. If we can really apply intelligent design to the details of nature, as its proponents suggest, we should find it to be a logical and sensible explanation for the history of life as well as for its current diversity and adaptability. Sadly, we can't. When we consider design in the light of the fossil record, we realize that our "designer" would actually be a "special creator," and that his work would be anything but simple and elegant. Instead he'd be a fitful and impulsive creator, building and destroying in countless ways that curiously resemble a supposedly nonexistent evolutionary process. When we apply design to the molecular basis of life, we find it supported only by a series of assertions that complex biochemical systems can be proved to be unevolvable. That "proof" is the assertion that systems composed of multiple parts lack function until all of their parts are assembled. Unfortunately, that assertion is simply not true—not even for the favorite examples of the ID supporters themselves, including the blood-clotting machinery and the bacterial flagellum. And, as we've seen, if science took the assertions of ID seriously, it would have closed off whole lines of research that have actually turned out to be productive.

Remarkably, design is an appealing idea only when we *don't* take it seriously. If all we're interested in doing is putting forth an alternative to evolution, no matter what the cost, design looks darned attractive. It's a one-word, one-size-fits-all explanation for every organism, every organ and structure and process in the world of life. And we can make it sound almost as scientific as evolution by linking it to information theory, biochemical complexity, and molecular biology. If all we really want is an alternative to evolution that has the look and feel of science, design might just work.

The real problems with design emerge when we indulge the idea, when we rev up the engines of human curiosity to see what design actually says about nature. For design to work, for it to succeed as a genuine scientific theory, it has to reward that curiosity. Unfortunately design is built upon a stunning lack of curiosity and a remarkable unwillingness to embrace scientific discovery. Design rests ultimately on the claim of ignorance, upon the hope that science cannot show evolution to be capable of producing complex organs, assemblies of molecules, or novel biological information. If evolution cannot achieve that, the argument goes, then design must be the answer.

Since any field of biology, including evolution, is filled with unsolved problems, intelligent design can be invoked as the default explanation for any one of them. But where does that explanation get us? As we have seen, not very far, and the reasons should be obvious. The hypothesis of design is compatible with any conceivable data, makes no testable predictions, and suggests no new avenues for research. As such it's a literal dead end, and seems intended to get us to do just one thing: to step back from science and acknowledge the creative efforts of an unnamed power behind the mysteries of life. As I will suggest, if that power exists, there is a much better way to pay it our scientific respects: namely, by solving those mysteries one by one.

FOUR

Darwin's Genome

EVERY NOW AND THEN an opinion research organization decides to ask Americans what they think of Darwin's theory of evolution. "Not much" is the usual answer. In fact Americans reject evolution in such numbers that non-Americans tend to shake their heads in disbelief that a scientifically advanced country could harbor such doubts about the central theory of biology. But harbor them we do. In the modern world America stands as the world champion of Darwin-rejection.

There is, however, a curious element to all of this: The answer to the evolution question depends on how it is framed, and whether it involves *us*. For example, a 2005 Harris poll[1] asked people a variety of questions relating to evolution. When human evolution was left out of the survey, the support for Darwin approached a majority. Asked if "all plants and animals evolved from other species or not," 49 percent of Americans agreed, actually giving evolution a slight edge over the 45 percent who disagreed. But when the same group was asked if *human beings* had evolved from an earlier species, the numbers changed dramatically. Now only 38 percent accepted evolution, while 54 percent rejected it. When the question was phrased in a way that involved the divine, only 22 percent maintained that we "evolved from an earlier species," whereas 62 percent said that "human beings were created directly by God." Clearly evolution loses significantly when respondents realize that we're talking about *our* species, and even more dramatically when the question seems to involve God.

Remarkably, the book of Genesis tells us that we humans were formed out of the dust of the earth, and evolution says pretty much the same thing—the only difference is in the details. But those details seem to matter, especially if they suggest that we have our ancestry in species that most of us would regard as "just animals." For many Americans this is the only aspect of evolution that really matters. In that respect it's worth noting that the Tennessee law under which biology teacher John Scopes was prosecuted in 1925 didn't actually forbid the teaching of evolution per se. Rather, it made it illegal to teach any theory of *human origins* that contradicted "the Divine Creation of man as taught in the Bible."[2] Evolution might be just fine to explain the ancestry of ferns and bluebirds and germs, but just keep those damn monkeys out of *our* family tree.

DARWIN'S GAMBLE

One doesn't have to be a Darwin scholar to realize that Charles knew he was taking an extraordinary gamble when he published *On the Origin of Species* in 1859. The reason, as Darwin clearly realized, was the implication his theory held for our origins. He was at first curiously reticent about saying anything directly implying that humans might be the direct product of evolution. Although his conclusion to *On the Origin of Species* promised readers that, as a result of evolution, "light will be thrown on the origin of man and his history," the book itself contained scarcely a word on our own species, leaving its readers to speculate whether its author actually intended to imply that evolution had given rise to humankind. Buoyed, perhaps, by the scientific successes of his theory, Darwin left no doubt about human ancestry when a decade later he penned *The Descent of Man.* If any of his readers had been willing to entertain the idea that the good Mr. Darwin might have intended to leave man out of his devilish scheme, this book made it clear that this reclusive biologist was determined to drag us right down into the muck of natural selection with the rest of creation.

Darwin's notebooks and letters indicate that one of the most compelling observations in the development of his theory was the lack of firm boundaries to the characteristics that define species. Individual species do have distinct characteristics, to be sure, for if they didn't, we would hardly be able to classify them. But to Darwin, a patient observer of the natural world, those characteristics revealed their own plasticity, their own tendencies to vary over time and space. Those variations were the raw material of evolution, and served to dash the idea that species were permanently fixed over time. Alfred Russel Wallace, the naturalist whose letters shocked Darwin into publication, had made exactly the same observation based on his fieldwork in Southeast Asia. The 1858 paper Wallace sent to Darwin had the revealing title "On the Tendency of Varieties to Depart Indefinitely from the Original Type," which, while not quite as catchy as "On the Origin of Species," was sufficiently descriptive to make the point—living organisms can change over time.

Did that include us? Did the struggle for existence include the human species, and if so, did it mean that our characteristics had likewise "departed indefinitely" from our distinctly animal ancestors? That was the question Darwin addressed in *The Descent of Man,* and his conclusions were clear:

> Man at the present day is liable, like every other animal, to multiform individual differences or slight variations, so no doubt were the early progenitors of man; the variations being formerly induced by the same general causes, and governed by the same general and complex laws as at present. As all animals tend to multiply beyond their means of subsistence, so it must have been with the pro-

> genitors of man; and this would inevitably lead to a struggle for existence and to natural selection.[3]

In other words the human species is affected by natural selection in the same way as any other organism. What that means, of course, is that evolution applies to us as well. It is in our future—and it certainly was in our past. We didn't appear out of thin air. Like every other species on this planet, we had a history. We evolved.

But if our species did have a history, where was it? Why hadn't the fossil ancestors that might document that history been discovered? Darwin didn't have an answer except to note that "in all the vertebrate classes the discovery of fossil remains has been a very slow and fortuitous process." In other words, just wait, and maybe we'd get lucky. But Darwin did a little more than simply note the absence of evidence. He actually told his readers where to find it.

> The fact that they [human progenitors] belonged to this stock [the old world monkeys] clearly shows that they inhabited the Old World; but not Australia nor any oceanic island, as we may infer from the laws of geographical distribution. In each great region of the world the living mammals are closely related to the extinct species of the same region. It is therefore probable that Africa was formerly inhabited by extinct apes closely allied to the gorilla and chimpanzee; and as these two species are now man's nearest allies, it is somewhat more probable that our early progenitors lived on the African continent than elsewhere.[4]

The boldness of this prediction, in the absence at that time of so much as a single prehuman fossil from that great continent, is stunning. But Darwin's confidence in his reasoning shines through in his careful prose, even a century and a half later. Investigators would, just as he predicted, find a treasure trove of prehuman fossils, and Africa would be the place to find them.

MISSING LINKS

Darwin's well-argued exploration of our animal nature and animal origins cried out for documentation in the form of hard evidence, and little by little that evidence emerged. Starting with a few fragmentary finds in the late nineteenth century and some remarkable discoveries in South Africa, human paleoanthropology developed into a distinct field by the middle of the twentieth century. Yet even as its practitioners zeroed in on eastern Africa and explored its fossil riches, public imagination remained fixed on the idea of a "missing link," a single fossil find that would tie us to our primate past and settle forever the issue of human origins. The irony of the missing link idea is that its dominance has made it remarkably easy for critics of evolution to ignore the fossil riches that were rapidly being accumulated and focus instead upon the notion that definitive ancestors would remain missing forever. They didn't, of course, and the last three decades have seen a remarkable explosion of fossil evidence that should have settled forever any question of our prehuman origins.

Foremost among this evidence is, of course, the famous Lucy fossil, a prehuman form that its discoverers enthusiastically placed at the very base of the human family tree. But the evidence also includes more than a dozen distinct species, each found somewhere in Africa and dating from the past 4 or 5 million years. In fact, the pace of discovery has now picked up to the point that the very notion of a single missing link is clearly out of date. We have, in reality, discovered so many missing links that the real question has become how to deal with this embarrassment of riches—in other words, how to connect the dots.

The richness of this evidence was brought home to me in a remarkable way a little over five years ago when an article in the scientific journal *Nature* described the discovery of a new fossil species in eastern Africa.[5] Dubbing their new find *Kenyanthropus platyops,* scientists tried to place it in perspective by producing a diagram showing how *K. platyops* might be related to other fossil species. When I read the article I did a double take at a figure that showed how the new discovery might fit into fourteen other previously discovered species in our recent ancestry. I was certain that I had seen that figure before—or something very much like it. But how could this be possible? The new fossil had just been discovered, and the research report with the figure had just appeared. So how could it be familiar? I spent the better part of a day fruitlessly searching for a solution and finally just gave up. Two days later the solution came to me: The reason the diagram seemed so familiar was that it was an almost exact match for the only figure that Charles Darwin had drawn for *On the Origin of Species* (figure 4.1).

Darwin, I suspect, worried that his ideas would be misinterpreted and took whatever steps he could to prevent that. His theory did not suggest a straight-line transformation of one species into another, but rather a spreading and branching process in which the accumulation of variation would lead to greater and greater species diversity. So, in his famous sketch he illustrated how one species would gradually develop greater diversity until it branched into two, and then branched yet again. Natural selection would prune away many of these branches, applying the unforgiving shears of extinction to most of them, until only a few of the divergent experiments remained. Over time a single species would give rise to a branching pattern of descent from which only a few new species emerged. This is, of course, exactly what the human fossil record now looks like, as the *Nature* illustration shows.

How do Darwin's modern critics respond to the flood of fossil intermediates that link us to prehuman primates? Very often they dismiss the wealth of fossil evidence as irrelevant. Those supposed intermediates? They're either clearly human or clearly apelike; there's nothing in between, and that's why evolutionary trees (like figure 4.1) of supposed prehuman ancestors just don't matter. An evolution enthusiast named Jim Foley has had great fun with this

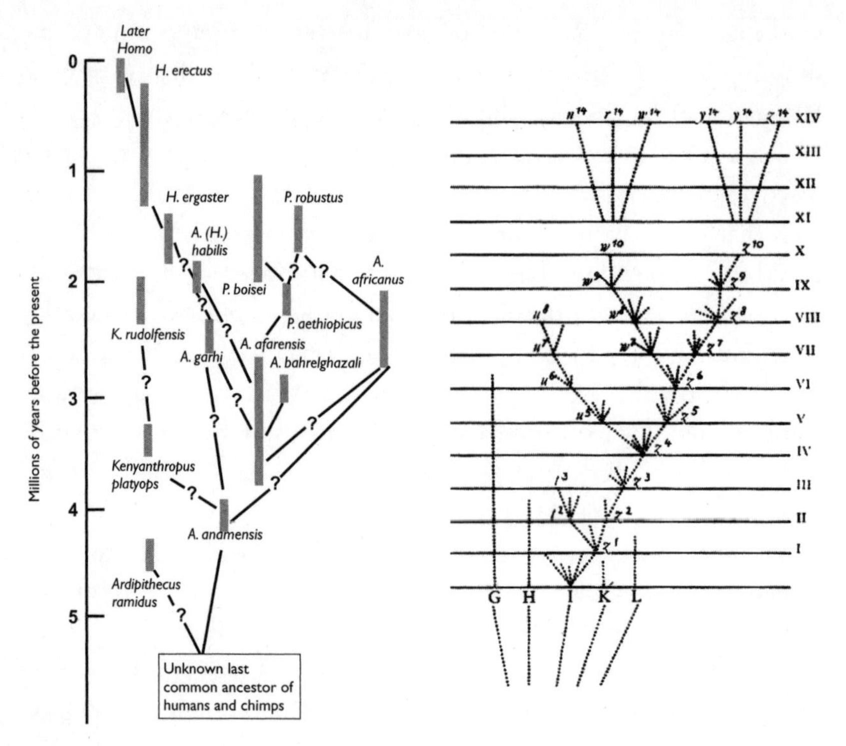

Figure 4.1: The human fossil record and Darwin's sketch from *On the Origin of Species*. A 2001 paper from the journal *Nature* included a summary diagram of human and prehuman fossil species *(left)*. Although the intent of the diagram was merely to compare a new find with previously known species, its branching pattern was a remarkable match for Darwin's illustration of speciation over time *(right)*, the only diagram he included in *On the Origin of Species*. (Left: *Kenneth R. Miller, drawn from data in D. E. Lieberman, "Another Face in Our Family Tree,"* Nature *410 [2001]: 419–20.* Right: *Charles R. Darwin,* On the Origin of Species, *6th ed., 1859 [New York: Oxford University Press, 1996].)*

contention. He took ten creationist articles and books, each of which disputed the human fossil record with this argument, and compared what they said about six specific prehuman fossils. As figure 4.2 shows, the writers all agreed that one of these fossils was definitely apelike, and two were definitively human. On the other three, they disagreed.[6]

Now, there's nothing unusual about disagreement over how to classify such fossils. Paleontologists differ on such matters all the

Specimen	Cuozzo	Gish (1985)	Mehlert	Bowden, Menton, Taylor, Gish (1979)	Baker, Taylor and Van Bebber	Taylor, Lubenow
ER 1813 (510 cc)	Ape	Ape	Ape	Ape	Ape	Ape
Java (940 cc)	Ape	Ape	Human	Ape	Ape	Human
Peking (915–1225 cc)	Ape	Ape	Human	Ape	Human	Human
ER 1470 (750 cc)	Ape	Ape	Ape	Human	Human	Human
ER 3733 (850 cc)	Ape	Human	Human	Human	Human	Human
WT 15000 (880 cc)	Ape	Human	Human	Human	Human	Human

Figure 4.2: Creationist classifications of hominid fossils. Six specific fossils are represented, with the brain size of each given in cubic centimeters (cc). Creationist authors are unanimous in insisting that there is a clear dividing line between human and nonhuman fossil primates. However, they disagree among themselves as to whether certain fossils should be found on one side of that line or the other. By showing how ten creationist publications deal with six specific fossil specimens, James Foley, who prepared this analysis, demonstrated the intermediate nature of several of these fossils. (Details of this chart, including references for each publication, are found in endnote 6 of this chapter.) *(Kenneth R. Miller, drawn from data provided by James Foley as part of the Talk Origins Web site.)*

time. But the nature of creationist dissent on these three fossils is delightfully revealing. Some of the evolution critics said that the fossils were clearly apes, while others flatly stated that they were human. Which group of creationists is right? I don't really know, and that's the point. In fact, I'm tempted to say they both are. What better proof could one offer of the transitional nature of the human fossil record than the profound lack of agreement of antievolutionists as to how to classify these fossils? Ironically, validation of our common ancestry with other primates comes directly from those who are most critical of the idea.

INSIDE STORY

Opponents of evolution often complain that Darwin's proponents receive substantial government funding, and that taxpayer support gives them an unfair advantage. In a sense they're right—but not in the way they think. In reality, evolution research (paleontology, speciation, natural selection, etc.) is one of the most poorly funded areas in all of biology. It's extremely difficult to get such work underwritten by government agencies, and many researchers are forced to compete for far more modest support from private foundations. Despite that tough reality, the last decade has seen a spectacular gathering of new evidence in favor of Darwin's great idea from a great, publicly supported effort—the Human Genome Project.

Each of us is born with a genetic inheritance carried on the forty-six chromosomes that make us human.[7] A typical human gene, the basic unit of inheritance, might contain anywhere from several hundred to several thousand DNA bases,[8] and some very large genes contain nearly a million. In the last decade of the twentieth century a major international project was launched to read and map the sequences of DNA bases in the human genome, and at the very beginning of the twenty-first century, that effort neared completion. Today, even though there are still a few rough spots, the DNA sequences of the human genome are a matter of record and are freely available from computer databases around the world.

So here's the relevant question for purposes of our discussion. Was that genome designed, or did it evolve? The critics of evolution like to say that the complexity of the genome makes it clear that it was designed, by which they seem to mean that a cosmic (or supernatural) designer sat down somewhere and typed out the A's, T's, G's, and C's of every human gene and chromosome. The beauty of that explanation, of course, is that it could, in principle, explain anything. We simply defer each and every question about the details of the genome to the designer, and leave it at that. Those six billion DNA bases are the recipe for building a human being, and they represent the complex, unique, direct, and inspired work of the designer. And very fine work it was.

But there's a problem with that analysis, and it's a serious one. The problem is the genome itself: It's not perfect. In fact, it's riddled with useless information, mistakes, and broken genes. And there's a more serious problem, too: It looks as if our genome was copied from somebody else's. The designer, apparently, is a plagiarist.

OUR MISSING VITAMIN

If you like orange juice as much as I do, you may drink it with a sense of both pleasure and responsibility. It tastes great, hence the pleasure. But it's also good for you, for lots of reasons, so you'd be justified in feeling virtuous every time you tossed down a glass. One of its benefits is that it provides vitamin C, a compound known to chemists as ascorbate. We need a certain amount of ascorbate in our diets because it plays a crucial role in the chemical reactions that help to build collagen, one of the body's most abundant and important proteins. Eat a diet with little or no ascorbate, and these reactions fail. Connective tissue, which helps to hold the body together, begins to break down from lack of proper collagen assembly, and the result is a disease called scurvy. Every grade school health class warns students about the dangers of vitamin C deficiency, and many remind them that the need for this nutrient was a discovery of the British navy as it sought ways to keep its sailors healthy. Eventually provisioners brought barrels of lemon or lime juice on board, leading to the nickname of "limeys" for British sailors. This much of the lore of vitamin C is common knowledge.

But have you ever thought of what other animals do to get their vitamin C? The answer, in most cases, is that they do nothing. Most mammals are able to make the vitamin C they need independently, a trick they pull off by means of a series of five enzymes in the liver that manufacture ascorbate from ordinary sugars. So what happened to us? It turns out that we have most of these enzymes in our livers, too, but we are lacking a critical one called gulonolactone oxidase, which biochemists call GLO.

Unless there was a plan to pump up sales of citrus fruits, why were we designed without a gene that could have made our dietary lives so much simpler? The people who believe our genome was indeed designed have heard these questions before, and they have a ready answer: Something can truly be designed and still be far from perfect, like a bad car or a very slow computer. Design, therefore, does not imply perfection, and a bad design is still a design. There's no reason to think that a designer had to make us metabolically perfect. Fair enough.

But the interesting part of the story is that we aren't exactly missing the GLO gene. In fact, it's right there on chromosome 8, in pretty much the same relative position in our genome where it is found in other mammals. The problem is that our copy of the GLO gene has accumulated so many mutations, in the form of changes in the DNA base sequence, that it no longer works. We've got to include vitamin C in our diets because we carry a defective version of our GLO gene. In effect, we all suffer from a genetic disease, which we can correct only by including vitamin C in our diets. What follows, of course, is a very logical question. If the designer wanted us to be dependent on vitamin C, why didn't he just leave out the GLO gene from the plan for our genome? Why is its corpse still there?

Arguing that a design doesn't have to be perfect doesn't explain why the designer would type out a gene with mistakes. Even really bad designers don't *intentionally* design radios with broken transistors or software that crashes every time you use it. ID proponents' answer in this case is more interesting: We must have a remnant of the GLO gene because the designer originally gave us a functional one, which became dysfunctional in the course of human history. The billions of bases in our genome must be copied every time a cell divides, and at some point in the human family a few copying mistakes popped up right in the middle of this gene. If our ancestors were living in a place where vitamin C was plentiful in their diets, the loss of the GLO gene really wouldn't have mattered, and over time it would have become fixed in the human population.

That sounds like a perfectly reasonable explanation, and so it is: All humans today carry those mistakes because all of us are the descendants of individuals who carried the same defects in their

GLO genes. Mistakes exist in this important gene, in other words, because they were inherited from a common ancestor. But in that simple conclusion lies the undoing of any claim for our separate ancestry as a species. We are not, you see, the only species in which the GLO gene is broken.

The need for vitamin C is also characteristic of a certain group of primates, the very ones that happen to be our closest evolutionary relatives. Orangutans, gorillas, and chimps require vitamin C, as do some other primates, such as macaques. But more distantly related primates, including those known as prosimians, have fully functional GLO genes. That means that the common ancestor in which the capacity to make vitamin C was originally lost wasn't a human, but a primate—an ancestor that, according to the advocates of intelligent design, we're not supposed to have. And there's the problem.

PLAGIARISM

Several times in my career as a college teacher I have caught students cheating on exams or term papers. Sometimes it's an easy thing to spot. They pass notes during exams, crib passages from the Internet, or copy from each other's papers. During a final exam in one of my courses several years ago I noticed two students carefully eyeing each other's work, and watched them long enough to be almost certain that one was copying an essay from the other. When the exam was over I set both papers aside for comparison and confirmed my suspicions. I called both students into my office the next day and told them that the similarities in their essays left no doubt that one had copied the essay from the other.

One of them challenged me to back up the charge. "Of course our essays are similar," he told me. "We're roommates and we study together. So naturally we gave pretty much the same answer to the same question. That doesn't mean we copied."

"No," I answered. "It doesn't." And indeed one of the students had gone to considerable lengths to conceal the similarity of his work to that of his roommate. He had used a different title, had switched the order of several paragraphs, and had clearly rewritten many of the sentences to make them read quite differently. "But I wasn't looking at superficial similarities. I was looking at something deeper," I explained, placing the paper in front of him and calling his attention to a series of words, each circled in red ink. "The two of you misspelled the same six words, and you misspelled each of them in exactly the same way." Immediately the students realized that I had them.

If the two students had worked independently, they might indeed have made similar points in a correct answer to an essay question. They might even have worded a few sentences in the same way. But the notion that the exact same spelling mistakes could crop up, independently, on two different papers was too much to give credence to. There is, after all, just one way to spell a word correctly, but there are an infinite number of ways to get it wrong. And when those errors match perfectly, there can be no doubt that those errors must have a common source. The students pleaded guilty, and threw themselves on the mercy of the court.

This notion of unique, matching errors is widely used to determine when one document has been copied from another. Mapmakers, for example, often protect their costly investment against another firm's copying by placing a mistake in the form of a nonexistent town or road in a place of little consequence. If an unscrupulous publisher decides to produce its own maps by merely copying their work, the original mapmaker can go to court and prove copyright violation by pointing out that the hapless plagiarists included the mistake in their map. Matching errors indicate a common source for both maps.

The same is true for genomes, and when the proponents of ID explain the errors and imperfections in our genomes as being the result of accumulated mistakes, they inadvertently make a wonderful argument for the evolutionary common ancestry of our species with other primates. Case in point: the human globin genes.

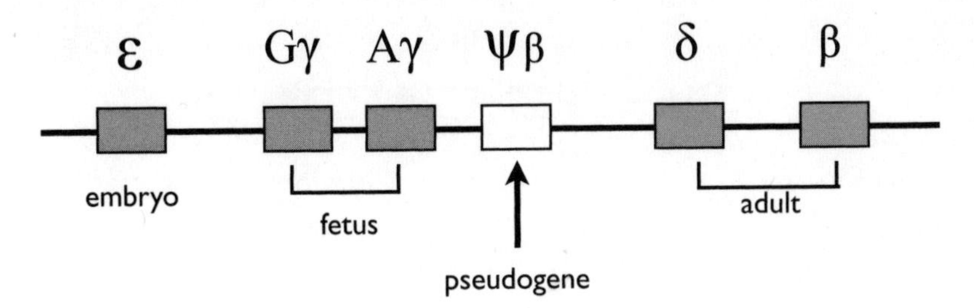

Figure 4.3: A simplified map of the human beta-globin gene cluster. This cluster, from human chromosome 16, contains five working copies of the gene, which are expressed at various times during embryonic and fetal development, as indicated. In the middle of the cluster is one nonworking pseudogene. *(Kenneth R. Miller.)*

Hemoglobin is the oxygen-carrying protein that makes blood red. A single molecule of hemoglobin consists of two copies of a molecule called alpha-globin and two of another called beta-globin. The genes for beta-globin, of which there are five functional copies, are found on human chromosome 16, and as shown in figure 4.3.[9] Right in the middle of this hardworking group of genes, however, is a broken one, which geneticists call a pseudogene. Its DNA base sequence is nearly identical to that of its neighbors, so it's easy to recognize it as beta-globin, but it contains a series of errors in its sequence that keeps it from working. Figure 4.4 shows the nature of these mistakes.[10] One of them prevents the gene from ever being copied into RNA (an essential step for a gene to work), one would prevent any RNA that did get made from directing the synthesis of a protein, and the remaining four would completely disrupt any protein that somehow managed to get produced anyway.

As we did for vitamin C, we might now ask how a designer could have allowed this to happen. And our ID friends would answer in the same way they did before: Our genome was originally designed with six working copies of the beta-globin genes, and therefore the loss of one of them from such molecular errors is no big deal. Therefore all humans alive today are descended from individuals in which those mistakes first cropped up. Fair enough. But guess what: We're not the only organisms with a set of beta-globin genes, and

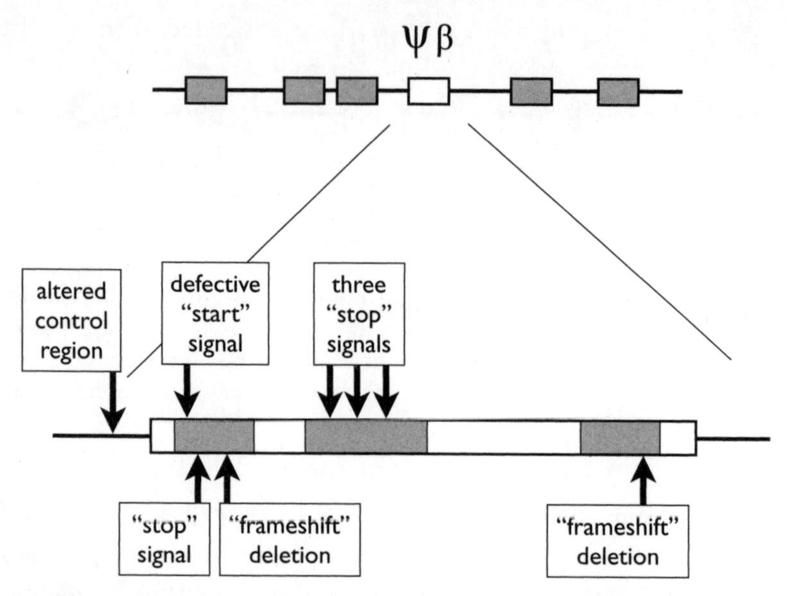

Figure 4.4: Genetic defects in the beta-globin pseudogene. A detailed analysis of the beta-globin pseudogene shows a series of mutations that have rendered it nonfunctional. The shaded areas within the pseudogene show exon regions, which would code for amino acids in beta-globin if the pseudogene were functional. Significantly, identical errors appear in the same pseudogene, in the same location, in the genomes of the chimpanzee and the gorilla. Such matching molecular errors indicate common ancestry for these three species. *(Kenneth R. Miller.)*

we're not the only ones with a pseudogene right in the middle of that set.

Gorillas and chimpanzees have them, too, and they are arranged in exactly the same way—five working copies surrounding a single pseudogene, just as in humans (figure 4.4)—but here's where life gets truly interesting. The gorilla and chimpanzee pseudogenes have *exactly* the same set of molecular errors. Just like the two kids in my class, they have matching mistakes. The student papers had a common source (the original essay), as do these two pseudogenes—the original version of the pseudogene in the ancestor of all three species.

There's no escaping the implication of these matching mistakes, and there's no point in arguing that six identical mistakes could have turned up independently in three unrelated species. The only sensible interpretation is that the original errors developed at random in a single common ancestor of these three species. In a court of genetic copyright law, any motion that a designer could claim originality for the human genome would be tossed out in a flash. Our genome is a copy, an altered one to be sure, but a copy, nonetheless, of earlier genomes that preserve even the mistakes our ancestors made in duplicating their DNA.

THE CASE OF THE MISSING CHROMOSOME

The cluster of beta-globin genes and our lost capacity to make vitamin C are hardly the only molecular errors that we share with our primate cousins. In fact, our genomes are packed with such little reminders of where we came from. Francis Collins, the director of the Human Genome Project, pointed this out in his wonderful book *The Language of Life,* and the examples he cites are worth noting.[11] Our genomes contain a surprisingly large number of base sequences that serve little function except to mark the point at which other pieces of DNA, some of them from viruses, jumped into and out of our genes. The molecular "scars" of these infections are everywhere, they're quite easy to recognize and catalog, they are passed along from generation to generation, and they are not the products of any sort of design. What do they tell us about evolution? As Collins points out, when one examines similar regions of the human and chimpanzee genomes, nearly all of these little pieces of molecular debris line up perfectly between the two species. There is only one possible explanation for such a match—they reflect events that occurred many millions of years ago, in the common ancestor of both species.

When I appeared as an expert witness in the 2005 federal trial regarding the teaching of intelligent design in Pennsylvania, I sought a quick way to bring the weight of the evidence for evolutionary common ancestry to the attention of the court. I began by explaining the matching errors in our beta-globin pseudogenes, and then read aloud from an issue of the journal *Nature* published just three weeks before the trial began:

> More than a century ago Darwin and Huxley posited that humans share recent common ancestors with the African great apes. Modern molecular studies have spectacularly confirmed this prediction and have refined the relationships, showing that the common chimpanzee (*Pan troglodytes*) and bonobo (*Pan paniscus*) are our closest living evolutionary relatives.[12]

But I wanted something even more compelling, something sufficiently graphic to stick in the mind of the judge and make it clear just how complete the scientific case for common ancestry really is. Just before the trial I found it. The question of our origins is, after all, a forensic one. Forensic analysis of crime scene evidence is an increasingly important part of scientific criminology. One doesn't have to be present during a crime to collect the evidence that remains, analyze it, and come to scientific conclusions regarding what happened and who made it happen—even for a crime that occurred years ago. So, why not apply the same approach to the origin of our species? All we need is a critical clue, and fortunately there is one sitting right in front of us: our chromosomes.

Evolution has produced a wealth of evidence—from fossils, genes, and physiology—that indicates that we share a common ancestor with the great apes, including gorillas, chimpanzees, and orangutans. But there's an interesting little chromosomal inconsistency in our story of common ancestry, and therein lies an opportunity to put evolution to the test. We humans have forty-six chromosomes, while all the other great apes have forty-eight. So how could we possibly share a common ancestry when we're apparently missing a couple of chromosomes?

Let's look at this question a little more closely. Our forty-six chromosomes are actually twenty-three pairs of chromosomes (since we inherit two complete sets, one from Mom and one from

Dad), which means that the great apes have twenty-four pairs. So, if we share common ancestry with these organisms, we humans must be missing a single pair of chromosomes. Could that chromosome pair have been lost in the line that gave rise to us? Not a chance. We know enough about primate genetics to understand that the loss of a complete chromosome pair (and all the genes they contain) would be fatal—to a human or a chimpanzee.

There is, in fact, just one way to explain the apparent absence of a pair of chromosomes in our species. In the line that led to us, two primate chromosomes must have been accidentally fused to form a single human chromosome. The beauty of this hypothesis is that it is testable. If one of our chromosomes was indeed produced this way, we ought to be able to scan the human genome and identify a chromosome with two halves, literally pasted together, from its primate ancestors. If we don't find such a chromosome, then the common evolutionary ancestry postulated for our species might be wrong. If, on the other hand, we do find such a chromosome, then we have once again found evidence that confirms evolution. Now all we need is a way to recognize that fusion and solve "the case of the missing chromosome."

ELEMENTARY, MY DEAR WATSON

Fortunately chromosomes themselves carry a series of landmarks that provide all of the necessary clues to search for a fusion event. The tips of chromosomes contain unique DNA sequences known as telomeres, which play a critical role in copying DNA. What would we expect if one chromosome were to become fused to another? As figure 4.5 indicates, we'd expect to find telomeres where they don't belong—right in the middle, at the fusion point between the two original chromosomes. Then, just to be sure, we could make use of another special feature of each chromosome called the centromere. Centromeres are the places where newly replicated chromosomes are attached to each other prior to cell division, and they have unique sequences of DNA bases that are easily recognized. If

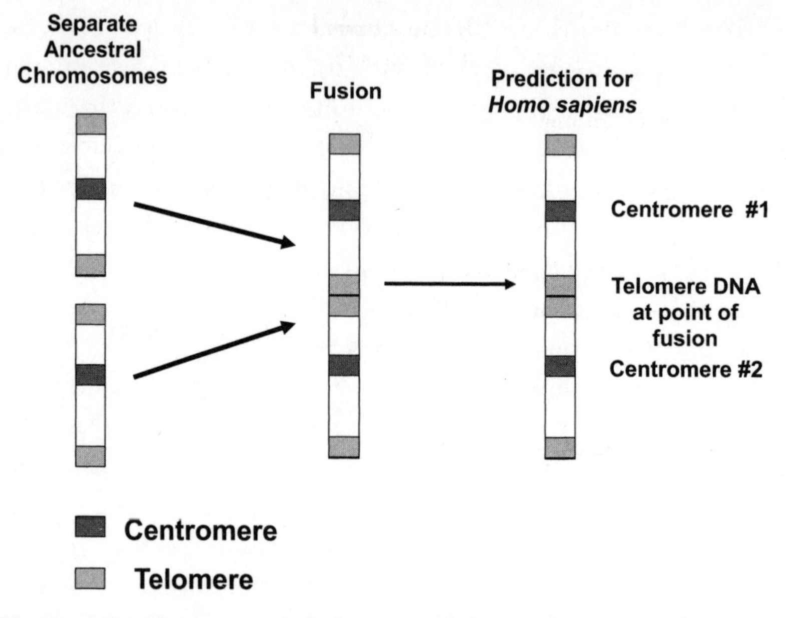

Figure 4.5: Chromosome fusion would leave distinct marks in the genome. An accidental fusion between two chromosomes could explain why humans possess forty-six chromosomes rather than forty-eight as do the great apes. However, such a fusion event would leave distinct marks in the new chromosome. Chromosomes contain recognizable regions at their tips known as telomeres, and other regions near their midpoints called centromeres. If two complete chromosomes fused together, telomere sequences would be expected to remain near the fusion site. In addition, the fused chromosome would be expected to carry two centromeres. *(Kenneth R. Miller.)*

our genome contained a fused chromosome, it ought to have two centromeres—and we might even be able to read the sequences on both of them and identify the sources of both centromeres. If evolution's ideas regarding our origins are incorrect, however, we should find no such chromosome in our genome.

Do we have such a chromosome? We do indeed: human chromosome 2. A 2005 scientific paper exploring two of the human chromosomes examined this issue in great detail. "Chromosome 2 is unique to the human lineage of evolution," its authors wrote, "having emerged as the result of head-to-head fusion of two... chromosomes that remained separate in other primates."[13]

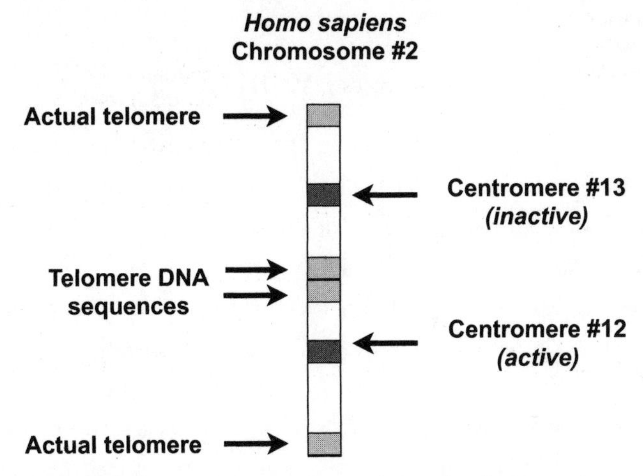

Figure 4.6: Chromosome 2 was formed by the fusion of two primate chromosomes. The second human chromosome displays each of the elements predicted for a fused chromosome. Telomere DNA sequences are found near the center of the chromosome, marking the precise location where two ancestral chromosomes were joined. In addition, as predicted, the chromosome contains two distinct centromere regions. One of these is inactive, which makes the chromosome more stable during cell division. The specific DNA sequences in each centromere closely match those of primate chromosomes 12 and 13, as noted. *(Kenneth R. Miller.)*

In fact, the molecular evidence for the fusion point is so strong that we can actually identify the exact region where the two chromosome tips were combined, where the two primate chromosomes were pasted together (figure 4.6).

Human chromosome 2 does indeed contain telomere DNA at its middle, at the fusion point, and it carries two centromere sequences that correspond to the centromeres from chimpanzee chromosomes 12 and 13. Furthermore the genes on human chromosome 2 are arranged in an almost exact match for the patterns of corresponding genes on the two chimp chromosomes. So clear is the match, in fact, that scientists working on the chimpanzee genome have now changed the numbering of chimp chromosomes 12 and 13 to chromosomes 2A and 2B, to match the human chromosome to which they correspond. The forensic case of the missing chromosome is settled beyond any doubt.

DESIGN'S NEW BURDEN

One of the great attractions of the ID explanation has clearly been the way in which it brushes the complexities of evolutionary change aside with the simple mantra "design." Can't come up with the exact details by which a cellular machine evolved? Then it must have been designed. Haven't figured out the source of a new gene? Maybe it was designed. Puzzled by the absence of apparent ancestors for an unusual species? Well, if it was designed, it wouldn't have any ancestors, would it? No reason to look for them! In every case design offers a neater, cleaner, and less troublesome solution. After all, when your explanation has no testable steps, there are no means to disprove it. It just sits there, almost like the smile on Alice's Cheshire cat.

But this easy preference for design as an explanation doesn't jibe with forensic analysis of the sort we've just discussed. If we find a criminal's fingerprints at the scene of the crime, we know that he was there. If he claims to have been on the other side of town at 3:00 P.M. but we have a security video of him entering the store at exactly that time, we can rule out his crosstown alibi. If we see him robbing a bank one day in New York and spending marked money from the same bank the next week in Florida, we don't have to prove exactly what happened on the six intervening days to know he was the thief.

The same logic applies to our studies of the human genome. The advocates of ID would like to insist that, unless we can reconstruct every step of the process by which evolution produced each feature of the human genome, then design remains a legitimate possibility. They will not allow, of course, any similar requirement for design theory, since the designer's intelligent work, by their definition, exists outside the laws of the natural world. But if we genuinely seek an answer to the question I posed earlier—namely, "Who wrote our genome?"—we can indeed find the solution by applying the same forensic methods used to solve crimes.

Whose fingerprints are on the broken glass, whose tools were used to break and enter? To get equivalent answers about the genome, we need only search through the clues left in the human genome itself and in those of other organisms. Is our genome, our arrangement of DNA bases and genes on chromosomes, a modified copy of earlier work, or is it an entirely new creation? Does a comparison with other organisms reveal common ancestry or unique, intelligent programming? In every case for which we have data—and that now includes our complete genome and the genomes of many of our closest animal relatives—the answer is clear. We're working with a modified copy, a genome loaded with inherited errors that has been shuffled and mutated and rearranged. We have, in short, a genome that evolved.

Remember the issue of a disproportionate amount of government funding being used to build up the case for evolution? As I wrote earlier, that's certainly true, but it's not because of money channeled specifically to evolutionary biology. Rather, it's due to the explosion of basic and applied research into the nature of genomes. In an age when detailed knowledge of the genome is at hand, the signature of evolution is everywhere.

Knowledge of the genome has placed a crushing burden on the claims of intelligent design. It can no longer just shrug at the details of living organisms and invoke a designer to account for them. It now has to explain why our genome everywhere proclaims "descent with modification," to use Darwin's favorite phrase for evolution. Can ID tell us why we carry a broken gene for vitamin C synthesis, or why the molecular errors that litter our genome match those of other primates? Of course not. Can it explain why one of our chromosomes looks for all the world as if it were pieced together from two primate chromosomes? Only if we suppose a designer with very strange intentions, including a plan to mislead and a determination to make things look as though they evolved. Mighty curious stuff.

Suddenly one can no longer simply wave at biological complexities and then ask, "How could this have happened?" The data of genes and genomes hold distinct patterns at the molecular level that design must also explain. The design interpretation works only if we pretend that we do not have these facts in hand, and it is threatened by each new detail, each new advance, each new discovery. It depends on a patchwork of coincidence and mystery, and seeks to describe the evolutionary consistency of modern genetics as nothing more than the unknowable whims of a designer's fancy. This is not the hallmark of a genuine scientific theory.

The beauty of evolutionary theory is that it is master of past and present, and that it ties each into a seamless fabric of science and of existence. Design, first and foremost, supposes a break between present and past: A special time when design events, which no longer occur, took place. A wall beyond which science cannot penetrate. A mystery world of wonder and magic. In effect, a magical, fairy tale past.

Life in the present demands more. The reality of evolution is in our genome, and it is the key to who we are.

FIVE

Life's Grand Design

IF I HAD TO GIVE a prize for the best idea that anyone in the antievolution movement has ever had, I'd award it to whomever came up with the term "intelligent design." Over the years the old standbys of "creationism" and "creation science" have served their purposes, but they've always had the serious weakness of revealing their religious nature too directly. If someone says he is a scientific creationist, there can be no doubt that he stands for something distinctly Bible-based. He might even claim, in order to fit the supposed chronology of Genesis, that the earth is less than ten thousand years old. The word "design," in contrast, doesn't sound religious and seems to take no position on the age of the earth, while the adjective "intelligent" appeals directly to our notions of purpose and meaning in the world around us. Indeed, no matter how you look at it, coining the phrase "intelligent design" has been a winning strategy for the antievolutionists.

We live in a country where nearly 90 percent of the people profess a belief in God, and I would argue that just about every one of them, myself included, believes that there is meaning to their lives, order in the universe, and a purpose to our existence. In that very distinct sense it can be said that a great majority of Americans believe in something like an "intelligent design" to the world around us. As a result the existence of arguments for design will command their immediate sympathy. After all, what's the alternative? Unintelligent design? Random design? Or no design at all? A universe of molecules weaving through space and time without meaning or purpose?

As I hope to make clear, those sympathies are misplaced. The choice of a name for the ID movement was deliberate and carefully calculated as a way to unify the opponents of Darwinian evolution and split its adherents, and it has managed to do both. As a label that defines a public relations effort, intelligent design has earned favorable comment from the president of the United States, cover-page status on major magazines, extended discussion on media talk shows, and acceptance by a large proportion of Americans. In political terms, it is a strategy that has worked, and worked well. But in scientific terms, it's been a nonstarter. Phillip Johnson, the retired Berkeley law professor who shaped the beginnings of the movement, has been remarkably candid on this point:

> I also don't think that there is really a theory of intelligent design at the present time to propose as a comparable alternative to the Darwinian theory, which is, whatever errors it might contain, a fully worked out scheme. There is no intelligent design theory that's comparable. Working out a positive theory is the job of the scientific people that we have affiliated with the movement. Some of them are quite convinced that it's doable, but that's for them to prove.... No product is ready for competition in the educational world.[1]

But if the scientists working on ID haven't managed to produce a "positive theory," then why are so many people eager to embrace it? Why is there nationwide pressure to include it in our schools if there's really no specific theory to teach? The reason should be clear: The very use of the word "design" implies purpose, order, and meaning to our existence. And what of those who argue against design? What of a scientific establishment that defends evolution from these frontal assaults on the central theory of biology? The tactical genius of the ID movement is that its very name has maneuvered the defenders of Darwin into a losing position, one in which they are made to seem as if they are arguing against meaning and purpose, to be claiming there is neither rhyme nor reason to the universe in which we live.

The great irony of this situation is that the defenders of science are well aware that that caricature is wrong. Scientists are, in fact, the very ones who search for order in our existence, who are best equipped to marvel at nature, and who find the deepest fulfillment in their exploration. Like others who have been lucky enough to make a career in science, I find the natural world filled with wonder and delight, and shake my head at those who would depict the scientific enterprise as something that would rob our existence of its meaning. Quite the contrary: The more we understand of nature, and the more thoroughly we investigate our past, the more deeply we can appreciate just how remarkable our lives truly are.

So I will come right out and say it. There is indeed a design to life. And that design arises out of the very fabric of the universe itself.

That may seem like a radical thing for a scientist, particularly a defender of evolution, to proclaim, but it's true nonetheless. It's a notion that stands firmly in the scientific mainstream, and it is anything but a concession to the "design" movement. It is, in fact, a direct contradiction of everything the ID movement stands for. There is indeed a grand design to life, and it's the very one first glimpsed more than a century ago by a fellow named Charles Darwin.

WHAT'S IN A NAME?

What's in a name? Everything. A rose may smell as sweet by any other name, but creationism apparently does not. In the 1980s efforts to require that schools teach something called "creation science" alongside evolution came crashing down, legal victims of the application of the First Amendment by federal courts. In 1987 the U.S. Supreme Court handed down a definitive ruling declaring that creationism and so-called creation science were inherently religious, and therefore had no place in public science education. The case was known as *Edwards v. Aguillard,* and at the time it seemed to have put an end to the creation science movement. Stephen Jay Gould, who had earlier testified as an expert witness against an Arkansas creation science law, observed: "We who have fought this battle for so many years were jubilant. The Court, by ruling so broadly and decisively, has ended the legal battle over creationism as a mandated subject in science classrooms."[2]

As I was to learn many years later, at the very same time that Gould was celebrating the Court's decision, a team of writers was pondering what to do with a manuscript they had produced on creation science. The book was to be published by the Texas Foundation for Thought and Ethics, and it was intended as a public school textbook to "supplement" the treatment of evolution in the existing biology curriculum. In fact, the first draft of the book, produced in 1983, was labeled *Creation Biology Textbook Supplement;* subsequent drafts were called *Biology and Creation* (1986) and *Biology and Origins* (1987). Early in 1987 its authors, Dean Kenyon and P. William Davis, settled on the title *Of Pandas and People: The Central Questions of Biological Origins,* which was used when the first edition appeared in print in 1989.

How had Kenyon and Davis reacted to the 1987 Supreme Court decision, a decision that seemingly dashed any hope of getting their project into classrooms? They fired up their word processors and revised their existing draft. We know about the history of this text because *Of Pandas and People* was at the center of the Dover, Pennsylvania, intelligent design trial in 2005. In response to a subpoena filed by the plaintiffs in the lawsuit, the publishers produced all of the early drafts of the manuscript, which revealed, in great detail, how its authors changed key words in the text. Barbara Forrest, a professor of philosophy at Southeastern Louisiana University,

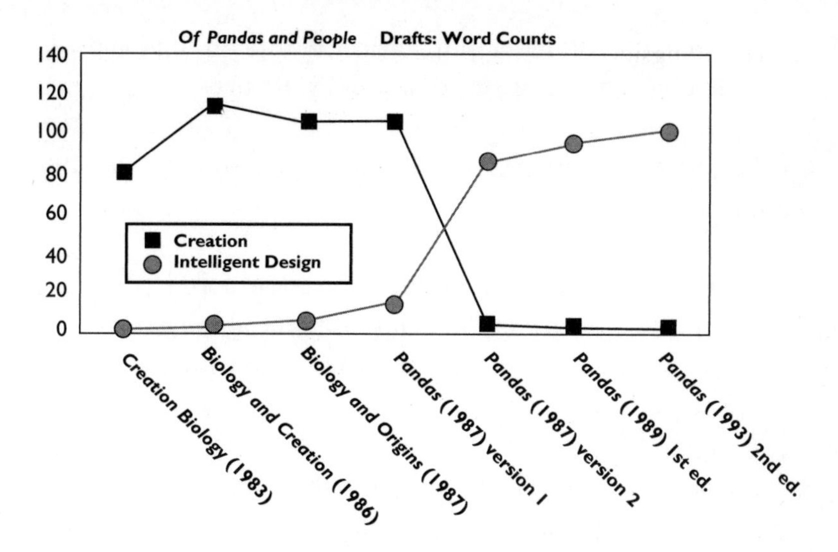

Figure 5.1: Intelligent design's relationship to creation science. An analysis of the frequency of words related to "creation" and to "intelligent design" in earlier versions of the manuscript for the intelligent design textbook *Of Pandas and People.* This analysis, prepared by Barbara Forrest for the *Kitzmiller v. Dover* trial, showed that intelligent design had been abruptly substituted for creationism in 1987. The change took place immediately after the U.S. Supreme Court branded creationism a religious doctrine. *(Kenneth R. Miller, redrawn from original provided by Dr. Barbara Forrest, Southeastern Louisiana University.)*

transfixed the courtroom during the Dover trial when she showed the results of these alterations (figure 5.1). Specifically, four early drafts of the book, the last one early in 1987, mentioned "creation" or "creationism" roughly a hundred times each. By contrast, "intelligent design" was mentioned less than a dozen times. Suddenly, in mid-1987, those numbers were reversed: "Creationism" fell toward zero, and "design" or "intelligent design" took its place.

A close examination of the manuscripts reveals exactly how this was accomplished. Here, for example, is the definition of "creation" in the early 1987 version:

> **Creation** means that the various forms of life began abruptly through an **intelligent creator,** with their

> distinctive features already intact—fish with fins and scales, birds with feathers, beaks, and wings, etc.

And here is the definition of "intelligent design" from the late 1987 version:

> **Intelligent Design** means that the various forms of life began abruptly through an **intelligent agency,** with their distinctive features already intact—fish with fins and scales, birds with feathers, beaks, and wings, etc.

This pattern of wholesale substitution shows that "intelligent design" was simply plugged in for "creation," while the "creator" morphed neatly into a "designer." In short, the authors changed the name of the supernatural agent responsible for the living world. The plan, quite clearly, was to circumvent the Supreme Court decision by producing a book that could still advocate "creation science" by calling it something else. That tactic was, as federal judge John E. Jones III would write in his 2005 decision in the Dover trial, an "astonishing" attempt to conceal the intent of *Pandas* by means of "a purposeful change of words . . . effected without any corresponding change in content. . . ."[3]

The new term, however, would catch on. In 1993 Phillip Johnson, who had earlier written a book highly critical of evolution, gathered together a group of like-minded individuals for a meeting in the coastal town of Pajaro Dunes, California. Present were the future leading lights of the ID movement, including Michael Behe, William Dembski, and Jonathan Wells. What emerged was a new strategy, built around the notion of "intelligent design," that would attack evolution by developing a distinct scientific alternative.

By changing the name of the rose, Johnson's disciples had made it possible to unify opposition to evolution. How so? Paul Nelson, a Discovery Institute fellow who also attended the Pajaro Dunes conference, explained this clearly in a 2002 article entitled "Life in the Big Tent: Traditional Creationism and the Intelligent Design Community."[4] Noting that the majority of dissenters from evolution up to the beginning of the ID movement had been "traditional" (young-earth) creationists, Nelson brimmed with satisfaction at how ID had now made it possible to draw together a "wider community" to battle evolution. By no longer relying explicitly on Genesis, ID had gathered Darwin dissenters of every sort under the banner of design, producing an alliance that could shake the foundations of evolutionary science. Once Darwin's ghost was banished, the victors could then battle over the "scientific narrative of design," including details such as the age of the earth, the nature of the fossil record, and the means by which the designer-creator acted. But for now, the strategy was to keep evolution squarely in the crosshairs and fire away.

The practical value of the "big tent" strategy cannot be underestimated. Traditional creationists, after all, rejected not just evolution, but nearly all of mainstream science. They quarreled with geology over the fossil record and the age of the earth, with astronomy over the distances to stars and galaxies, with cosmologists over the age and origin of the universe, and even with physicists over the laws of thermodynamics. In the eyes of most Americans, who appreciate the value of science, these quarrels made the young-earth creationists look ridiculous, and limited their public acceptance. Their attacks also helped unify the disparate elements of the scientific community, since nearly all of their disciplines were targets. At a stroke the formation of the ID movement set all of these controversies aside, not because it necessarily accepted the elements of mainstream science, but because of its careful and deliberate focus on a single target—on Darwin. The goal was to divide and conquer: Win the battle against evolution, and other sciences would come later.

THE MATTER OF DESIGN

So, what of the true design of life? We live in a material world. In many ways that realization is at the very heart of science itself. By seeking material, or natural, explanations for what we see and experience, science has changed the world—or at least our view of it. We no longer look for gods to pull the sun across the sky, or evil spirits to explain our daughter's illness. Physicists have split the atom, then split its fundamental particles, and then looked even deeper into the textures of material existence. Astronomers have lifted the veil of distance from our tiny neighborhood and peered so deeply into space that we now approach the very beginnings of the existence of the universe.

Modern biology arises out of a conviction of the material nature of life and would be impossible without the infusion of the physical and chemical sciences into the study of living things. Indeed, it was not so long ago that scientists had to argue to their colleagues, with some passion, that the secrets of life were indeed to be found in the matter of which living things are composed. Erwin Schrödinger's marvelous 1944 book *What Is Life?* is a perfect case in point:

> As we shall presently see, incredibly small groups of atoms, much too small to display exact statistical laws, do play a dominating role in the very orderly and lawful events within a living organism. They have control of the observable large-scale features which the organism acquires in the course of its development, they determine important characteristics of its functioning; and in all this very sharp and very strict biological laws are displayed.[5]

Schrödinger, one of the pioneers of modern physics, was seeking to bring the same intellectual rigor to biology that had driven the twentieth-century revolution in the physical sciences. The key, as he argued throughout his short book, was to treat life as a chemical and physical phenomenon, and he was right. James Watson, codiscoverer of the double helical structure of DNA, was later to remark that reading Schrödinger's book inspired him to seek a chemical explanation for the nature of the gene. DNA itself, of course, was that explanation. Today, therefore, it is commonplace to state what only a century ago would have been regarded with doubt and suspicion: namely, that life is a chemical and physical phenomenon.

This realization is a more radical break with the past than many of us realize today. For most of human history it was commonplace to believe that living and nonliving matter were truly different. The notion that ordinary matter could support the wonders of life, whether in a dandelion or a sparrow, seemed so inconceivable that many were willing to believe in an élan vital, or vital force,[6] that animated living matter. In the early nineteenth century the first attempts to apply the infant science of chemistry to living matter were called "organic" chemistry, reflecting a view that the chemistry of living things would somehow be fundamentally different from ordinary, or inorganic chemistry. Today, of course, chemists no longer hold this view, and the term "organic chemistry" survives with a quite different meaning, to denote the study of the chemistry of carbon-based compounds and their derivatives, whether associated with living matter or not.

How does all of this apply to the issue of evolution? What it means, first and foremost, is that the capacity for life is built into matter. In fact, the key molecules of life are largely constructed from just a few relatively common atoms, such as hydrogen, oxygen, carbon, nitrogen, phosphorus, and sulfur. In that sense the chemical properties of these atoms are what makes life possible, and if any of these properties differed in a significant way, life would be quite different—or might even fail to exist at all.

Physical scientists, in particular, have marveled at the remarkable precision with which the fundamental constants of nature must be honored in order to make our universe—and our lives—possible. Nowhere has this phenomenon been dealt with as elegantly as in Martin Rees's wonderful book *Just Six Numbers*.[7] Rees, who holds the title of Astronomer Royal at Cambridge University, constructed his narrative by choosing six numerical constants and deftly demonstrating how slight changes in any of them would have made life as we know it literally impossible.

Some of Rees's technical arguments are beyond the scope of this book, but a close look at one of them will serve to make the point. Under the right conditions (very high temperature and pressure) two protons and two neutrons come together in a reaction known as fusion to form the nucleus of an atom of helium. Since the helium nucleus itself has two neutrons and two protons, at first glance this doesn't seem very remarkable; four particles are joined to make a nucleus that consists of, well, four particles. But there's more to the story. The fused helium nucleus weighs just a little bit less than the weights of the two protons and two neutrons from which it was formed—0.7 percent less, in fact. What happened to the lost mass? It was released in the form of energy by the fusion process, and it is directly related to the strength of the force that holds protons and neutrons together in the atomic nucleus. That 0.7 percent doesn't sound like much, but it amounts to a tremendous release of energy—nothing less than the thermonuclear fusion that powers the sun and provides the explosive energy of a hydrogen bomb.

Now, what would our universe be like if this number had been just a little bit bigger or a little bit smaller? If it were bigger (maybe 0.8), the nuclear force would have been just a bit greater. Such a change would have made fusion occur much more easily than it does today. Would that have been a good thing? Not at all. The result would have been scores of runaway fusion reactions that would have drained the young universe of its hydrogen atoms, and no stars like our sun, no solar system, no earth, no water, and no life would have formed. But if the force had been just a little bit weaker, say 0.6 percent, it would not have been strong enough to hold protons and neutrons together. In this case fusion would never have occurred, and the universe would consist of nothing but hydrogen atoms. An interesting universe, one capable of generating the chemical diversity that makes life possible, has to get these numbers just right. Ours, as luck would have it, has done exactly that.

To Rees these six numbers are not merely a list of constants to be written inside the back cover of a physics textbook. They are nothing less than the recipe for building our universe, in all its complexity and grandeur, and their implications are extraordinary. One might conclude, for example, that the age and vastness of the universe renders our recent appearance on this small and out-of-the-way planet as nothing more than an insignificant afterthought in the grand scheme of things. And yet, as Rees makes clear, it's nothing of the sort. In reality a universe in which the evolution of creatures like human beings is even possible has to be precisely as vast and as old as the one in which we actually live. If the universe were smaller or younger, we simply wouldn't be here. It is scientifically true, as Walt Whitman once observed, that "a leaf of grass is no less than the journey-work of the stars."[8]

But maybe it isn't luck. Maybe the fortunate coincidence of fundamental constants in our universe is proof of a higher power, a cosmic architect whose careful planning ensured that we would live in a place where life was not only possible, but absolutely certain. This is the so-called anthropic principle, the observation that the universe seems to be structured in a way that makes human life possible. Some would then argue that it must have been constructed with that very possibility in mind.[9] Here, surely, is scientific proof of the existence of an intelligent creator.

Maybe so. But there's a deep logical problem at the heart of the anthropic principle, and it's been apparent since the idea was first suggested in the 1970s. Taking as a starting point the observation that you and I are alive, at least in the immediate present, it's obvious that we must live in a universe where life is possible. If we didn't, we wouldn't be here to talk about it. So, in a certain sense the fact that we live in a life-friendly universe merits little more than a big "Duh." *Of course* we live in a universe where the six big numbers make life possible.... Where else could we live? The anthropic principle doesn't prove the existence of a higher power or a higher intelligence. All it proves is that we are indeed alive—something we knew long before anyone thought to attach a fancy name like "anthropic" to the "principle."

But Rees is nonetheless taken with the sheer delicacy of the "recipe" for our universe, and finds that it demands an explanation. His favored one is the notion of the multiverse, an idea suggesting that ours is only one of many possible universes, each sprouting in its own direction from the primordial stuff of existence. If separate universes, each with different combinations of constants, were spun off in this way, it's hardly surprising that one (or a few) of them would have the conditions that make life possible. And that, of course, is the universe in which we live.

There's plenty to be written about here, and plenty has been written. Indeed, there's almost a cottage industry among cosmologists and physicists dedicated to explaining and exploring the implications of these remarkable ideas. It is not my purpose to add to this fine work, except to make one modest observation. However one accounts for the nature of our universe, however one explains the perfection of its recipe of constants, one fact remains: We live in a universe where the conditions that make life possible are built into the material fabric of existence. The ways in which matter and energy behave in our world are not only consistent with life, but are absolutely necessary for it. In effect the architecture of our universe, from the structure of a grain of sand to the unimaginable expanses of galaxies, is a blueprint for life—maybe, even, for human life.

A MACHINERY OF CHANGE

If the universe is made for life, what can we say of life itself? Is natural history a record of the sudden emergence of our living world in a single burst of creative energy? Or does it document something else, something suggestive of a process of change and adaptation akin to what happens in the world today? The answer, of course, is the latter. The first living organisms appeared on this planet roughly 3.5 billion years ago, and had we been there at the time, there wouldn't have been much to see. Life was microscopic for more than 2 billion years, producing nothing more than the sorts of microbes that today we call bacteria. Only in the last billion years did things get interesting, as the first cells with nuclei appeared and, a few hundred million years later, the first organisms made up of hundreds and even thousands of cells.

Life changed, and many of those changes were dramatic. The first true animals appeared around 650 million years ago, and barely a hundred million years later the earliest representatives of major animal groups, such as the arthropods (now including insects and crustaceans) and the chordates (now including mammals and reptiles), came on the scene. The geological period in which this occurred is known as the Cambrian, and it was so dramatic a flourishing that many paleontologists compare it to an explosion. The Cambrian was actually not an explosion at all, but a 30-million-year period of body plan experimentation and diversification that led to most of the major groups of animals alive today. But even the richly populated world of the Cambrian was not today's world. There were no insects, no modern fish, no land plants, no flowers, no birds, and not a reptile, a dandelion, or a frog. All of that was to come.

In the more than 500 million years that followed, the first reptiles, birds, and mammals, as well as the first flowering plants, came on the scene. There were great extinctions, and in their wake great diversifications. Life probed, explored, and colonized. Once lit, the spark of life spread everywhere on this great blue planet and has never ceased its endless process of change and adaptation.

What does the reality of natural history say with respect to the notion of design? First, it tells us that the specific details of today's living organisms were not the direct product of a flash of design in the dim and distant past. If they had been, life wouldn't be about change; it would be about stasis. The work of the designer would have been manifest in the perfect balances of nature, in the enduring constancy of his creations. Instead natural history reminds us that nothing lasts forever, and that the living landscape of the planet is forever changing. Second, it tells us that the ability to adapt to change and even to generate it is one of the most striking characteristics of living things. The design of life, ironically, includes an ability to change its own design.

How could that possibly be? Aren't organisms designed to perpetuate themselves faithfully from one generation to the next, carefully duplicating their genetic heritage and passing it along to their offspring? Well, sort of. But in reality no organism has yet to be identified that copies its genetic material without making a few mistakes, or mutations, in that information, and there are good reasons for that. As Frank Rothman, the marvelous professor at Brown University who first taught me biochemistry, was fond of pointing out, no organism ever copies its genetic information perfectly, for if it did achieve such "perfection," it would quickly find itself at an evolutionary dead end. Unable to adapt to changing conditions or to new competition, before long it would be driven to extinction by a host of more flexible competitors. As a result evolution itself, driven by natural selection, favors organisms that are able to change.

Rothman's commonsense insight made me think twice about what we dare to call "errors" or "mistakes" in biology. If apparent perfection in DNA replication is actually a handicap, then perhaps we should reconsider what we mean by the word "error." One might say that those chance mistakes aren't mistakes at all in the deeper sense, but the rich raw material of natural selection, the variations that make life possible. This much biologists have known for more than a hundred years. But today there is something genuinely new on the horizon. For the first time biologists are beginning to understand exactly how that change happens, and how it is channeled in ways that evolution can use.

This new understanding comes from the rapid advances that molecular biology has made in comprehending the machinery of the cell and in unlocking the mysteries of development. For the very first time we are beginning to learn how a living system can at once be incredibly complex and at the same time incredibly tolerant of change.

This has always been difficult to appreciate, because in our ordinary experience, complex systems are brittle. Simple machines are reliable, while complicated ones contain so many interdependent parts that they tend to fail in unexpected ways. If you doubt this, open the hood of your car and switch a couple of wires in its engine, or reverse a few lines of code in your computer's operating system. It is highly unlikely that you'll make things better, and quite likely that you'll make a royal mess of things. At first one might expect the same of life, given the biochemical complexity of the cell and the informational complexity of the genome. How could "random" evolutionary change ever reach into the staggering machinery of life and change things for the better?

How indeed? As puzzling as that question has always been, it has begun to yield to a pair of great discoveries that lie at the very heart of life on earth. The first of these is that living cells share an incredible amount of their basic biochemical machinery. Almost twenty years ago, when biologists began to investigate the ways in which cells controlled their own growth and division, they discovered a series of key regulatory proteins. These proteins interact with one another and with other proteins to determine the timing of essential events, such as when a cell will copy its genetic information and when it will divide to form two new cells. Interfere with these processes, and all hell breaks loose. That much one might have expected.

But what very few scientists anticipated was what came next. These key proteins turned out to be almost identical in "model systems" studied in the lab, such as sea urchins and yeast, and in humans. In fact human proteins could be substituted for similar proteins in yeast, and the machinery still worked. The basic core processes of life are the same across all kingdoms of life, from microorganisms to plants to animals. How, then, are we different from yeast? They key, as has been suggested by several researchers, is the flexible reuse of the same core components in different ways. What happened in the first two and a half billion years of life on earth, according to this analysis, is that these core components were selected and refined for adaptability, and that from this selection came the organisms that survived and prevailed in the struggle for existence—ourselves included.

This view is brilliantly described in a recent book, *The Plausibility of Life,* by Marc Kirschner and John Gerhart.[10] These two well-known experimental biologists argue that living organisms are biochemical machines that don't merely produce variation, but *facilitate* it in a wide variety of ways. The conservation of core components and processes among all living cells makes it possible to encourage mutation and genetic changes in ways that have the potential to benefit the organism:

> Central to our argument is that these [core] processes, many of which have been conserved for hundreds of millions or even billions of years, have very special characteristics that facilitate evolutionary change. They have been conserved, we suggest, not merely because change in them would be lethal (although that might be a factor), but because they have repeatedly facilitated certain kinds of changes around them.[11]

In other words, one of the reasons that these particular core processes have been preserved in evolution is that they provide a basic framework around which beneficial evolutionary change is possible. At its most basic level, the design of life doesn't resist change, but welcomes it.

> Instead of a brittle system, where every genetic change is either lethal or produces a rare improvement in fitness, we have a system where many genetic changes are tolerated with small phenotypic consequences, and where

> others have selective advantages, but are also tolerated because physiological adaptability suppresses lethality.[12]

Kirschner and Gerhart make the claim that theirs is a "major new scientific theory," and while this may be stretching it just a bit, their point is still compelling. Life has evolved in a way that favors change, is prepared to tolerate it, and channels it into useful genetic variation—exactly the sort of stuff that natural selection can go to work on.

REWRITING THE PLAN

If the concept of facilitated variation has begun to revolutionize our understanding of evolution at the cellular level, developments at a somewhat higher level are even more exciting. For centuries biologists who studied development gazed in wonder as the embryos of many animal species went through their intricate choreographies of growth, movement, and change to become the fully formed organisms that emerged from egg or womb or cocoon. Indeed, when I studied development in college, embryology was largely a descriptive science, punctuated by an experiment here or there hinting at the genetic and biochemical controls that surely lay just beyond our understanding.

In the last quarter century all of that has changed. One could attribute this progress to any number of researchers, but my first choice would be Ed Lewis, an unassuming biologist who worked at the California Institute of Technology on the genetics of *Drosophila* (fruit fly) development. Lewis had been trained in the honored tradition of fruit fly genetics established by the first great American geneticist, Thomas Hunt Morgan, and understood the significant advantages of the fly as an experimental system. *Drosophila* is easily bred in the lab, has a short generation time, and produces plenty of offspring, all of which make it ideal for genetic analysis. The fly, like many other animals, develops its body through a process of segmentation. Its segments, more than a dozen in number, are most easily seen in the fly's larvae (commonly called maggots). When the larval stage is finished, a pupa is formed, and after a few days an adult fly emerges from it. Earlier researchers had shown that each one of the segments in the larva formed a particular part of the adult fly's body, and Lewis's interest was in identifying the genes that told a particular segment what to become in the adult.

Lewis discovered that a series of genes, now known as the Hox genes, determine the fates of each segment of the developing fly. One gene was turned on in the very front segment, another in the next, and still another in the next. Interfere with these genes, by design or by accident, and what might emerge was a fly with an extra set of wings, or with feet dangling from where its antennae or mouth parts should be. Lewis's clever experiments demonstrated that while hundreds of genes might determine the detailed nature of each of these body parts, only a few actually call the tune. At once, the Hox genes were elevated to the status of master control genes, guiding the process of development.

But there were two unexpected surprises in this little group of genes. The first was almost too convenient to be true: The genes were found on a single chromosome, a single piece of DNA in the fly's genome, and they were arranged in the *exact* order in which they were expressed in the fly itself. This extraordinary lineup took biologists by surprise, but it was nothing compared with the next shock. The Hox genes weren't confined to flies or even to insects as a group but were found in virtually all animals, including mammals. In fact, the same pattern, the same head-to-tail lineup is found in humans as well. We have four clusters of the Hox genes, each on a different chromosome. Each gene in these clusters corresponds to a particular segment of the developing human body, and each one of them is a close match to a Hox gene in the fly.

The implications of these fundamental discoveries are still being felt in basic research, but they have already changed the way we think about animal development. They have, in fact, opened up an entirely new field of inquiry casually called evo-devo, a combination of the study of development and the study of evolution. The discovery of a small set of master genes that control development, of which the Hox genes are just a part, led biologists to realize that the recipe for building the animal body is controlled by remarkably few genes—and that by studying small changes in the recipe, they could also study how these genes produce *variation,* the raw material for evolution. In effect animals possess a tool kit for generating body form, and, incredibly, it is the same kit whether that animal is a honeybee, a fish, or an elephant. Developmental biologist Sean Carroll puts it this way in his book *Endless Forms Most Beautiful:*

> First of all, this is entirely new and profound evidence for one of Darwin's most important ideas—the descent of all forms from one (or a few) common ancestors. The shared genetic tool kit for development reveals deep connections between animal groups that were not at all appreciated from their dramatically different morphologies.
>
> Second, the discovery that organs and structures that were long viewed as independent analogous inventions of different animals, such as eyes, hearts, and limbs, have common genetic ingredients controlling their formation has forced a complete change in our picture of how complex structures arise.[13]

Carroll's first point adds to the groundswell of genetic evidence for evolution, but his second point is especially telling. Because the tool kit allows for remarkable flexibility within the modular, segment-by-segment organization of the body, it establishes a plan that, in effect, is optimized for evolutionary change. The master control genes provide a framework through which evolutionary change can be channeled.

> Why are existing body parts and genes the more frequent pathway to innovation? This is a matter of probability. Variation in existing structures and genes is more likely to arise than are new structures and genes, and this variation

> is therefore more abundant for selection to act upon. As François Jacob explained so eloquently, Nature works as a tinkerer with available materials, not as an engineer does by design. The invention of wings never occurred from scratch, but by modifying a gill branch (insects) or forelimbs (three times). Trends in evolution reflect the paths that are most available and therefore those taken most frequently.[14]

In Carroll's view, what happened in the millions of years preceding the Cambrian period was the gradual construction of this tool kit. The earth's first multicellular organisms, which appeared in the Ediacarian period as much as 100 million years before the Cambrian, had to evolve a way to deal with all those cells. It took millions of years, but eventually they developed a system of switches that could activate alternate programs of development in different parts of the body. This enabled them to lay down an anterior-to-posterior axis, and then to switch on one program of development to produce the head, another for limbs, another for the abdomen, and another for the tail. This revolutionary development was so powerful that it led to the explosion of body plan experimentation that characterized the Cambrian period, and the successful experiments gave rise to the major animal groups that persist to this very day.

LESSONS FROM AN EMBRYO

As Neil Shubin of the University of Chicago has explained, evolution doesn't act directly on the bodies of animals. Rather, it acts on the recipe that builds those bodies, which gives it far more flexibility than even Darwin might have imagined. By tinkering with a common developmental mechanism, variations in the way in which the recipe plays out can produce organisms as different as tuna fish, parakeets, and puppy dogs. The residue of these actions is especially clear in the common legacy of development that we can see in embryonic development.

More than a hundred years ago Ernst Haeckel, a German biologist, identified extensive similarities in the early stages of embryonic development in a number of vertebrates. Haeckel, an accomplished artist and illustrator as well as scientist, produced a series of drawings to highlight these similarities, drawings that have been widely reproduced and copied in textbooks right up to the present day. Struck by a number of common elements in development, including the fact that all vertebrates, even those that lack gills, develop gill-like slits as embryos, he proposed a "biogenetic law" that the embryonic development of an organism would reveal its evolutionary history. Haeckel's "law" was widely cited as evidence for evolution, even as biologists found it riddled with exceptions and contradictions, and his drawings turned out to be less than accurate. In particular he had exaggerated the similarities and downplayed the differences between several embryos in an overzealous effort to support his "law."

In recent years opponents of evolution have used Haeckel's supposedly "fraudulent" drawings as evidence against evolution, or at least as evidence of the degree to which evolutionists will lie and misrepresent to support Darwinism.[15] While it is indeed correct that the differences between vertebrate embryos are greater than Haeckel's drawings reflected, today we can ask a deeper and more fundamental question. We can look beyond the superficial similarities of shape and size, and examine the actual genes expressed in each part of the embryo during development. When we do this, the results are even more striking than Haeckel's own conclusions (figure 5.2).

As modern developmental biology shows, the morphological similarities that so transfixed Haeckel actually understate the evolutionary case. Each of these embryos possesses the same developmental tool kit, revealing both their common ancestry and a similarity of form and function produced by the workings of the evolutionary process. Today the vertebrate embryo not only argues for the reality of evolution, it also helps to show us in detail how that process works.

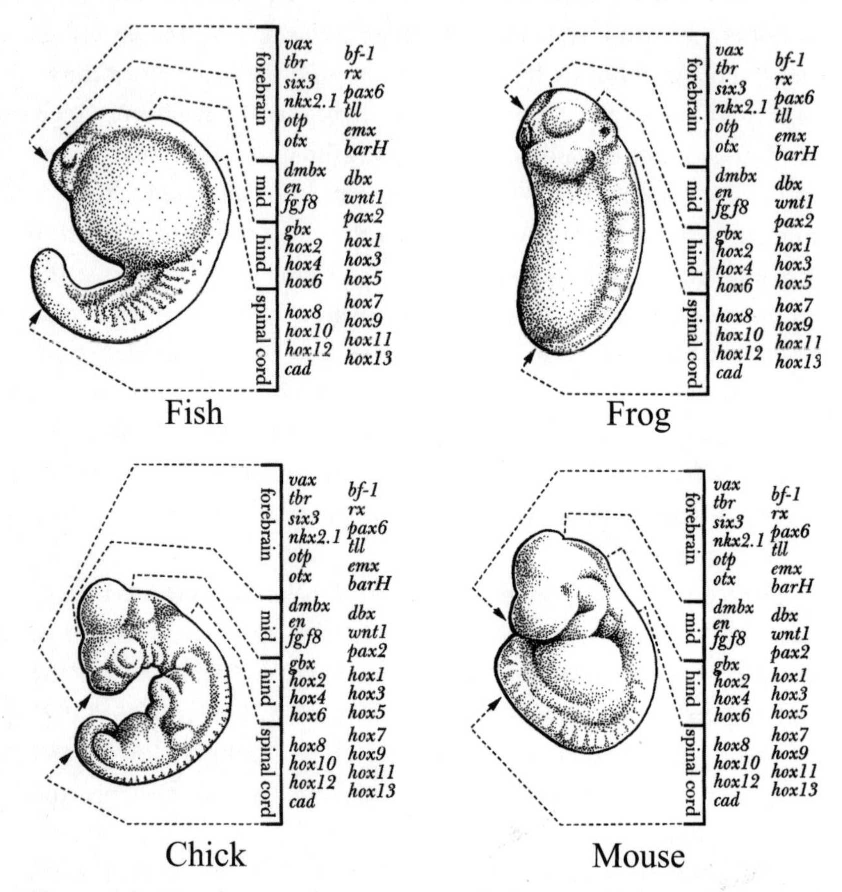

Figure 5.2: Vertebrate embryos compared. For years biologists have noted that embryos from different vertebrate classes show certain striking similarities as they develop. Critics of evolution have instead preferred to emphasize the distinct differences between such embryos. A deeper comparison can now be made with the tools of molecular genetics. A genetic map of the nervous system compartments in which master control genes are active shows that the very same sets of genes are expressed in corresponding compartments of each type of embryo. (Tissues surrounding the chick and mouse embryos have been removed for clarity.) *(Drawn by John Norton and used by permission. From M. W. Kirschner and J. C. Gerhart,* The Plausibility of Life *[New Haven: Yale University Press, 2005].)*

The lessons of these discoveries are profound and far-reaching. At one level they enable us to understand evolution at a level of detail never before possible. Carroll has argued that we now no longer need to make a distinction between the two types of change known as macroevolution and microevolution. He argues that evo-devo has shown that the two types of change are not really so different after all:

> The continuity of the tool kit and the continuity of structures throughout this vast time illustrate that we need not invoke very rare or special mechanisms to explain large-scale change. The extrapolation from small-scale variation to large-scale evolution is well justified. In evolutionary parlance, Evo Devo reveals that macroevolution is the product of microevolution writ large.[16]

To the ID crowd those are chilling words. They suggest that we really do know enough about the mechanism of evolutionary change to account for the large-scale changes that produce genuine novelty. By suggesting that the capacity for macroevolutionary change is already present in life, they do away with the need for a designer to infuse such change into it. In short they state that we need not look to forces above nature to account for the living world, and that the scientific case for intelligent design has failed. The nature of life is such that these extraordinary appeals are simply not necessary. The wonder of life is that the capacity for changing its own recipe is built right into it.

THE GRAND DESIGN OF LIFE

We have reached a point in science where the very phenomena that used to make us feel small—the vastness of the universe, our brief existence in time's long history, our apparent irrelevance to nature—can now be understood in a way that makes us the centerpiece of existence. We know that our very presence demands a universe of this size and age, and that only a precise mix of fundamental constants could have brought all of this about. We may indeed feel small when we contemplate the nighttime sky, but today we recognize that our very being requires that exactly such a universe be spread out before us in all its stirring beauty.

Life on earth is not a stranger to this universe, not a curious exception to its cold and heartless physical laws. Life is, rather, its most remarkable feature, its glory, maybe even its purpose. Life is built upon the physics and chemistry of matter itself. The reality is that we live in a universe in which the possibility for life is contingent upon the laws of nature and as such is woven into the very fabric of existence. By any standard we live in a universe that is simply brimming with evolutionary possibilities.

Our planet is one place where those possibilities have come to fruition. Not only did life emerge on this planet, but life mastered it. It adapted to conditions on the earth and then it changed them. It altered the gases in the atmosphere, reshaped the chemistry of the oceans, and remade the land forever. Life explored the possibilities of existence on Planet Earth, and evolution drove that exploration. Evolution's winners were the organisms that most successfully found a pathway to change and as a result have come to dominate existence on our planet. We are among those organisms, but we possess something special, something that no other organism to our knowledge has ever had—the ability to see and to understand how we came to be. That is a precious gift, and we must never lose it.

Evolution really does tell us something deep and profound about the world in which we live—something that Darwin glimpsed but that is much more obvious today. As it turns out, there really is a design to life, but it's not the clumsy, interventionist one in which life is an artificial injection into nature, a contradiction of its physical laws. Rather, it is a design in which life emerges from the laws of the universe around us. That conclusion is unavoidable, robust, and scientific. The elegant universe is a universe of life. And the name of the grand design of life is evolution.

SIX

The World That Knew We Were Coming

PUBLIC DISCUSSIONS OF EVOLUTION provoke strong feelings, and anyone who takes the side of Darwin in public had better be prepared to encounter them. Over the years I've gotten used to hostile questions that inevitably follow my talks to some audiences. In the summer of 2000, at the very end of a long question-and-answer session during my brief tour of Kansas, I was confronted by a question I'd heard many times before: "How can you tell me that I'm just an animal? How can you say that I'm no better than the beasts? That the only things that matter in life are to struggle, survive, and mate? There's just got to be more to life than that." Having voiced those fearful sentiments, the member of the audience sat down. That was all the argument he felt the need to make against evolution.

The depth and reality of the fear he expressed, deep and primal, cannot be doubted. And, as I will argue, it would be a terrible mistake to glibly attribute its existence, as many of my colleagues do, to nothing more than scientific ignorance. The truth is that evolution strikes at the heart of what and who we are. Are we of this world? Or are we of another? Are we angel or beast? Did we arise in a flash of divine inspiration or did we crawl slowly out of the muck? For too many people, evolution answers each and every one of these questions in *exactly* the wrong way. It tells us, to put it almost too plainly, that we are not special, that we have become nothing more than Darwin's beasts, that human nature has been reduced from the heavenly to the mundane.

In the parking lot that night, another person, perhaps just a little too shy to confront me in front of the crowd, asked that question again. "How can you sleep at night," he asked, "having written textbooks telling students they're nothing more than animals?" Why didn't I have the courage, the common decency, to admit the awful damage that evolution does to what young people think of themselves?

ARE WE REALLY JUST ANIMALS?

Of all the questions a skeptical person might ask about Darwin, I've come to believe that this is the most important one. Does evolution mean that we are nothing more than beasts?

Unlike questions about the age of the earth or the transitional fossils that link mammals to their reptilian ancestors, this is not the kind of question that scientists can easily answer, and there's a good reason for that. It's simply not a scientific question. To many of my scientific colleagues, that means that it's not a question worth answering.

But they're wrong. In some ways, it's the only question about Darwin's work that really matters.

When I had said my last good-bye that night, the heat had broken just enough to let me roll down the window on my rented pickup truck. As the moon crawled above the horizon, the rush of cool air along I-70 made me forget some of the events of the day and turn back to memories of another night, long ago. Once again the moon was rising in the darkness, and once again I was enjoying a break from summer heat. But on that night one of my daughters, no more than four years old at the time, sat by my side, sharing my delight at the night sky. Or so I dared to hope. Then she broke the silence with a question I have remembered ever since. "Daddy?" she whispered. "What are people for?"

It was hardly a scientific question, but it was no less worthy of an answer. In fact, the memory of my daughter's remarkable inquiry, nearly twenty years before my nighttime drive across the prairie, affected me in exactly the same way as those last few challenges put to me in Manhattan. I had been ready with an answer to both questions. But were they the right answers? How deeply had I actually looked into Darwin's mirror, how completely had I accepted my own ancestry from the world of beasts? Wasn't there a twinge of concern, a final reservation, a last reservoir of doubt from which I had secretly made peace with myself and tried to salvage a few fragments of that special sense of uniqueness that had vanished forever with the publication of *The Origin*?

In a very direct and personal way, I have written this book to confront some of the troubling questions that surround our understandings of human origins. My goal is to explore both our ancestries and our anxieties, to probe our connections with the natural world, and finally to see if we can make sense of them. Darwin's story of evolution speaks to all of us, and tells us something profound, welcome or unwelcome, about human nature. What we all must ask, ultimately, is whether the true story of human origins is to be feared and hidden, or embraced and celebrated.

Although I have encountered more than a few individuals who would argue, at least as an intellectual exercise, that any sense of meaning and purpose to life is mere illusion, I have a hard time believing them. I don't think their hearts are in such claims, in part because I can see them leading lives of such drive and conviction that their own purposes seem abundantly clear. Deep inside the inner lives of all of us, I suspect, there is something that provides the drive to propel us forward, something that resonates with sentiments like those found in Max Ehrmann's poem "Desiderata":

> *Beyond a wholesome discipline,*
> *be gentle with yourself.*

You are a child of the universe
no less than the trees and the stars;
you have a right to be here.
And whether or not it is clear to you,
no doubt the universe is unfolding as it should.

The words "You are a child of the universe... you have a right to be here" convey a ringing certainty that speaks to the human need to be sure of the world around us. "No doubt the universe is unfolding as it should" assures us that, however chaotic and disorderly the events of our lives may seem at the moment, we should take heart, for we were meant to be. And, in the eyes of many, that's exactly the problem with evolution—it says that we *weren't* meant to be and that the way things are unfolding isn't part of anybody's plan or purpose.

Many of my well-meaning scientific colleagues dismiss contemporary opposition to evolution as the hopeless product of biblical literalism. They figure that critics who take the Bible as a literal scientific treatise are so badly off course that there's just no helping them, so why even bother to introduce them to modern ideas in geology or astronomy, let alone evolutionary biology? But that view misses something even more unsettling among those who feel threatened by Darwin's great idea. The motivating power of this unsettling idea is far greater than the simple fear that the Genesis account might be figurative and not literal. It is the deep, raw fear that we might not be the result of a preordained pattern of history, that the universe might not have us in mind after all, and that we were indeed *not* meant to be here. The worry is that the universe is not "unfolding as it should" but, rather, that it's just unfolding.

BY ACCIDENT?

Today's anti-Darwin literature is rife with descriptions of "random evolution" and the "chance results of natural selection." Advocates of creationism and ID word their attacks on evolution so as to make it seem that the principal purpose of teaching evolution in our schools is to demoralize our young people by telling them that their lives are without meaning. Former Pennsylvania senator Rick Santorum, an outspoken antievolutionist, expressed this point of view in a national radio interview in 2005:

> It [evolution] has huge consequences for society. I mean, it's where we come from. Does man have a purpose? Is there a purpose for our lives? Or are we just simply, you know, the result of chance? If we are the result of chance, if we're simply a mistake of nature, then that puts a different moral demand on us—in fact, it doesn't put a moral demand on us—than if in fact we are a creation of a being that has moral demands.[1]

Santorum expressed a concern common to almost all antievolutionists—namely, that there is a link between an element of chance at the core of evolution and a host of bad consequences for society. At its simplest this means, Santorum claimed, that if we are indeed a "mistake of nature," there are no moral demands on us, and if there are no moral demands, then the standard of human behavior will fall to its lowest level.

After a spate of school murders in 2006, the creationist ministry Answers in Genesis posted this statement on its Web site:

> As a society, we reap the consequences of the unquestioned acceptance of the belief in evolution every day. It diminishes your worth and reduces human beings from being "made in the image of God" to being mere players in the game of survival of the fittest.[2]

The Answers in Genesis statement echoed the earlier assertions of former congressman Tom DeLay, who linked the killings at Columbine High School in Colorado directly to the science of Charles

Darwin. We should expect such behavior, according to DeLay, so long as our young people are taught that they are "nothing more than glorified apes evolutionized out of a sea of muck."[3]

Such notions of accidentality, of randomness, are much more than a rhetorical device intended to put evolution at odds with the notion of a divine purpose. They are instead an attempt to argue, by careful use of terminology, that our own lives will become pointless, disconnected, and meaningless if we dare to accept what scientists have been telling us about evolution, and that society as a whole will fall apart as a result.

It is only fair, therefore, to ask a question that goes to the heart of this analysis: Is the connection that evolution's critics so easily make between the operation of "random chance" in a material world and a lack of meaning valid?

We might begin by asking what is really meant by "random." One might say that "anything" can happen in a random event, but this is not really true. For example, most people would quickly agree that the winner of tonight's state lottery drawing will be picked "at random," but that's not the same thing as saying that *anyone* can win. I know for certain, for example, that I'm not going to win—because I haven't bought a ticket. The winner may be drawn "randomly," but in this case that means picked from a well-defined population (the ticket holders) and not from all possible individuals who might be happy to receive a prize. So, if we can indeed apply the word "random" to a lottery pick, it means that we can surely use it to refer to an *unpredictable* outcome chosen from a limited number of possibilities.

In a lottery drawing the forces that constrain those possible outcomes are obvious. The randomly picked lottery winner is limited to the set of individuals who hold tickets. In evolution the constraints may be less obvious, but they are still there. Although we often speak of mutations, changes in the genetic material, as being "random," that does not mean that all conceivable changes are equally possible or equally likely. The range of genetic changes is limited by the chemistry of DNA itself, by existing interactions within the cell and by preexisting constraints in the process of development. They are also quickly subjected to the forces of natural selection, which are most definitely nonrandom. In an organism that depends upon camouflage color or streamlined shape for survival, changes that interfere with either would soon disappear. Changes that enhanced either would tend to persist and become more common. Natural selection tends to drive evolution in the directions it favors, and as such is far from a random process.

Like tonight's lottery pick, it would be more accurate therefore to describe the nature of genetic change as "unpredictable" instead of "random." We may expect that natural selection will favor the evolution of protective coloration in a certain species that is prey for a larger one, and still not be able to predict what form that camouflage will ultimately take. We may even know that our own selection of chromosomes was determined in an unpredictable process in the primary sexual organs of our parents. When sperm and egg cells are formed, the twenty-three pairs of human chromosomes predictably separate from each other in a process known as meiosis. Although there are rare (and sometimes unfortunate) exceptions to the way in which this process works, generally each reproductive cell will get a single representative of each pair of parental chromosomes.[4] What is truly unpredictable is which member of each pair of chromosomes a particular sperm or egg cell may get. Ask the parents of multiple children just how different those kids can turn out, despite drawing their chromosomes from the same "pool" in their parents, and you can add personal testimony to this little refresher in high school biology.

The example of meiosis is particularly telling because it directly addresses the issue of individual origins. My own individual genome, the collection of chromosomes that guides my growth and development, that gave me blue eyes and brown hair and blood type O, was decided by a series of "random" events. These include the outcomes of meiosis in the reproductive cells of my mother and father, and then the chance meeting of just one sperm cell in several million with an egg cell.[5] The same is true for every human being—indeed, nearly every living thing—on this planet.[6] And yet it is common for people of faith to say that their own being was preordained, that God called them by name and has a plan for their existence. The notion that the personal existence of each individual human being was willed by God is written into the moral theology of many religions and serves as the basis for codes of personal behavior in many of them. Nonetheless, I have never heard a sermon preached against the evils of genetics, or on the corrupting effects of meiosis on our youth.

Somehow we have managed to reconcile the randomness of meiosis and fertilization with a sense of purpose and value in our lives, despite the inherent unpredictability of the process. One might suggest that this is because preachers and religious moralists have never heard of or simply don't understand meiosis, but I'd like to give them a little more credit than that. The real reason, I suspect, is that the constraints within which meiosis operates are so obvious and clear. These boundary conditions, these constraints, are absolutely necessary to the successful completion of meiosis, and even though its specific outcome is truly uncertain, it is easy for just about anyone to understand that the process is indeed predictable in a more general sense. Therefore, we can rationalize the notion of purpose and intention into a random process, just as long as that process operates within certain boundary conditions and achieves an understandable outcome.

So why do the random and unpredictable aspects of evolution provoke such a different reaction? Why is it so often argued that if random evolution is true, human existence is nothing more than a cosmic roll of the dice, and our actions are without meaning or consequence? Why can we deal so easily with meiosis and not with evolution? The reason, I think, is obvious: Evolution has been described in a way that makes it impossible to appreciate the constraints within which it operates. Unlike meiosis, evolution is presented, by critics and proponents alike, as a truly random process in which all outcomes are equally probable. Therefore our own existence carries no more meaning or significance than a toss of the cosmic dice.

If that were indeed true, the critics of Darwin would have a telling point in their public struggles against evolution, however mistaken they might be on matters of science. As we've seen, however, they are quite wrong about the science. In the remainder of this chapter, I will show why they are wrong about this issue, too.

CONTINGENCY

Any process dependent on a sequence of events is said to be "contingent," and evolution is most definitely a contingent process. Change just one or two of the events in a contingent sequence, and you may get a completely different outcome. The jargon of contingency may be confusing at first, but one of the easiest ways to understand it is to see how it applies to a much more familiar process—the game of baseball.

Full disclosure first. I love the game. I played it for countless hours as a youth, and became involved in the version of it known as fast-pitch softball when my two girls were young. After years of coaching I became an umpire, and I have umpired the sport at the collegiate and even professional levels.[7] Baseball, like any sport, involves arguments with officials, and I've been in more than my share. As you may know, occasionally these arguments result in a formal protest by one of the teams, and a game is then played out "under protest." By rule, protests cannot involve issues of judgment. I may have indeed called a strike on a low pitch, or missed seeing a ball hit the chalk line in the outfield, but these are issues of judgment and cannot be protested.

Applications of playing rules, however, *can* be protested. For example, let's suppose that a play occurred in which the bases were loaded with one out, and a deep fly ball was hit to center field. The runners prepared to tag up in anticipation of the catch, but the runners on second and third both left their bases before the catch. The catch is made (second out), the runner on third scores easily, but the runner who started from second is thrown out at the plate for the third out of the inning. One run scores, and the inning is over. But the team in the field now tosses the ball to third base and asks the umpire to declare that runner out, too, for leaving the base early. The umpire agrees that the runner did so, so there is no issue of judgment, but declares that since we already have three outs, the inning is over and the run scores. That's a mistake in applying the rules. The umpire should have allowed the team in the field to get a "fourth out" and take the run off the board. This mistaken ruling actually occurred in a game that one of my colleagues officiated, and a protest was filed. It was upheld, because the umpire had indeed made a serious error.

How does contingency apply in such a case? In the remedy for the successful protest. You might think that the league would take that run away and adjust the score. Not so. Because the gain or loss of a run affects every element of strategy and play that follows in a ball game, the only remedy for a successful protest is to go back to the field, make the correct ruling, and restart the game from that point on. And that's exactly what's done. The umpire who makes such a mistake has to return to the field and work the rest of the continued game without pay, and in some associations he even loses his original game fee—so there's a strong incentive to get it right in the first place!

The principle is clear: Change just one event in a contingent sequence of events—a ball, a strike, a runner called out—and everything that follows is affected. Any historical process, anything that involves a linked series of events, whether a baseball game, a nation's history, a military campaign, or evolution itself, is also a contingent process.

The contingent nature of evolution was brilliantly explored by the late Stephen Jay Gould in his book *Wonderful Life.* Taking his title from a classic Christmas movie, Gould reminds his readers of a point in that film where the character George Bailey (played by James Stewart) wishes out loud that he was never born. George's guardian angel then decides to show him what life would have been like if he had never existed, and George discovers that his local community and the lives of people in it would have been remarkably different—and not for the better. George comes to realize that he has indeed had a "wonderful life" in terms of his impact on scores of people, and he decides to make the best of things despite his problems. As movie fans know, things do work out in the end for Mr. Bailey, but Gould's point is not the happy Christmas ending. Rather, it's the impact that one apparently insignificant individual can have on the stream of events that follows.

Gould's scientific focus in *Wonderful Life* is the explosion of animal life that occurred during the Cambrian period more than 500 million years ago. Paleontologists such as Simon Conway-Morris had recently shown that some of the Cambrian fossils may actually represent our own ancestors (and those of all vertebrates), and Gould questioned whether the ultimate success of these organisms was a sure thing. He asked his readers to imagine natural history as a videotape and to rewind it into the Cambrian. If we let the tape run again, he wondered, would we get the same outcome? Absolutely not, was his conclusion, and I think he was quite right. Natural history, just like baseball or politics, is filled with contingent events and need not inevitably play out the same way. A second running of the tape might involve different mutations, different extinctions, and different cosmic events. Maybe the asteroid that wiped out the dinosaurs would miss the earth this time, and mammals would continue to live in the bushes and shadows, never giving rise to primates or to humans. Maybe the ancestors of vertebrates would be driven to a Cambrian extinction, and mollusks or echinoderms would come to dominate animal life on earth. It's almost certain that it would be a different life the second time around.

According to Gould, the contingent nature of evolution ensures that no two runnings of the tape of life would ever come out the same way. This reflects the unpredictable nature of the evolutionary process that so many critics of Darwin find so disturbing. And Gould's message to them is that they are right to be worried:

> By taking the Darwinian "cold bath," and staring a factual reality in the face, we can finally abandon the cardinal false hope of the ages—that factual nature can specify the meaning of our life by validating our inherent superiority, or by proving that evolution exists to generate us as the summit of life's purpose.[8]

According to Gould, "factual nature," or the scientific view of life, cannot demonstrate either our superiority as a species or our position at the purposeful "summit" of living things. You cannot look back at the history of our planet and detect a predetermined pathway leading straight to us. And therein lies the problem with any view of life that sees evolution as God's means and method of creation. As Gould emphasizes, we know enough about evolution to understand that it is a truly contingent material process, and no creator could have been sufficiently sure of its outcome to use it to produce a specific creature—like us. We should take the "cold bath" and get over it.

As compelling as Gould's narrative may be, his view of intention and outcome is exceptionally narrow. The only question he really asks is whether this exact world of life would emerge in a second running of the videotape of natural history, but it begs a much larger question. In a sense, it's almost like rerunning the videotape of one's own life to a point prior to the events of meiosis and fertilization within your parents, and asking if you would reappear at birth exactly as you did. Once again, the answer is no, but it's clearly a very different kind of no. I'd be a different person with a different genome, but I'd still have inherited my chromosomes from the same two parents. In a sense the same constraints that applied to my genetic makeup the first time would still apply the second time, and I would still be my parents' child.

The issue at hand is what sorts of comparable constraints might apply to our evolutionary videotape. While the contingent events of natural history would be different, and the outcome would be different as well, the organisms that emerged would still be the children of life on earth. They would still be subject to the specific laws of chemistry and physics that define our universe, and would still be the products of evolution and natural selection. How strong are the constraints imposed by those conditions? To what degree do they guide and determine the course of evolution? To what extent do they make the emergence of organisms like us a sure thing? Those are the questions that Gould's analysis overlooks, and here my answer is completely different from his. In a sense, I don't find the Darwinian "bath" to be chilly at all.

CONVERGENCE

There are all sorts of ways to live in the natural world. You can do very nicely by being small and asexual and simply absorbing your food, as many bacteria do. But being absolutely enormous, very sexy, and photosynthetic works, too, as any oak tree will tell you. The demands of existence require that organisms develop adaptations to fit their environments, but those adaptations can be remarkably different. Just as some humans work with their hands, some with their minds, and some not at all, organisms can fit into an almost unimaginable variety of "positions" on the evolutionary playing field. To represent this variety, the American scientist Sewall Wright invented the concept of *adaptive space,* a complex space in which small niches exist, representing every type of adaptation. All true adaptive spaces are multidimensional, which makes them impossible to draw, but one might imagine a three-dimensional space in which the three axes of the system represent size (large to small), shape (extended to streamlined), and speed of movement (from no movement at all to very fast). Placing various organisms into this three-dimensional graphic system, you'd find that slow-moving organisms, whether large or small, can be just about any shape. Large organisms that move quickly, however, tend to be streamlined, and this is particularly true of large, fast-moving animals that live in the water.

The specific demands of regions in adaptive space apply even if organisms are not closely related to one another. One might take as an example three large, fast-moving aquatic animals: a dolphin, a tuna, and an ichthyosaur (an extinct swimming reptile from the period of the dinosaurs). Each of these belongs to a completely different branch of the vertebrate evolutionary tree, and yet each of them has a remarkably similar shape. The reasons for this convergence of shapes aren't difficult to figure out. The laws of hydrodynamics apply equally to all organisms living in water, and evolution produced these fast-moving animals of similar size and shape due to an identical set of physical constraints imposed through natural selection.

As Simon Conway-Morris, the paleontologist of Cambrian fossil fame, has pointed out, convergence is one of evolution's most striking features. He finds Gould's failure to take convergence seriously a shortcoming that places his conclusions about the nature of evolutionary change on very thin ice. Stepping back and looking at the course of evolution over time, one can see that evolution essentially "explores" any given adaptive space. It pokes and probes the multidimensional landscape of adaptation, and when two or more organisms arrive in a particular niche on that landscape, they are subject to the same physical constraints. The way each adapts to those constraints can produce striking similarities among unrelated organisms, and examples abound.

Roughly 100 million years ago the continent of Australia became separated from the continental landmasses of Southeast Asia, an event that literally cut off Australia's mammal population from the rest of the world. At that time marsupial mammals (which give birth to underdeveloped offspring and carry them in pouches) and eutherian mammals (which produce more developmentally advanced young) lived side by side over much of the planet. As luck would have it, only marsupials survived in the newly isolated landmass of Australia, while eutherian mammals came to dominate the rest of the world.[9]

Competition between the two forms led to the extinction of nearly all marsupials outside Australia, while within it marsupial mammals had the run of the place. The results offer a spectacular lesson in convergence. In mainland Eurasia, Africa, and the Americas mammals filled a variety of well-defined niches in adaptive space. They produced hunter-predator species like wolves and foxes, burrowing animals like the common mole, and tree-climbers like the flying squirrel. Naturally the same niches in adaptive space were available in Australia, but only marsupials were present to take advantage of them. In effect the same videotape of mammalian evolution was waiting to be run, not at a separate time, but in a separate, isolated location. What course did mammalian evolution take on the island continent? It produced hunter-predator marsupials like the dingo and the Tasmanian tiger, burrowing marsupial moles *(Notoryctes)*, and even a marsupial version of the flying squirrel *(Petaurus breviceps)*.

From mouse to mole, from fox to rabbit, evolution's Australian second act filled the very same niches with marsupials that it had filled elsewhere with placental mammals. Could it be that the course of evolution is not quite so "random" after all?

The sources of evolutionary convergence of Australian marsupials with mammals on other continents are neither mystical nor mysterious. Evolution explores adaptive space, and similar types of adaptive space were found in both Australia and on the continental mainland. Although evolution found its way to all those spaces, it took quite different routes to get there in each one. The animals of Australia, as both visitors and natives will tell you, are indeed unique. So, when evolution got two parallel chances to fill the same places in adaptive space, it did not produce exactly the same results. Score one for Gould. But the general forms that evolution has produced in Australia, even given a different starting group of ancestral mammals, are remarkably like those that emerged on the continental mainland. Score one for Conway-Morris.

In fact, as Conway-Morris has pointed out in detail,[10] organisms arrive repeatedly at the same evolutionary solutions for a wide

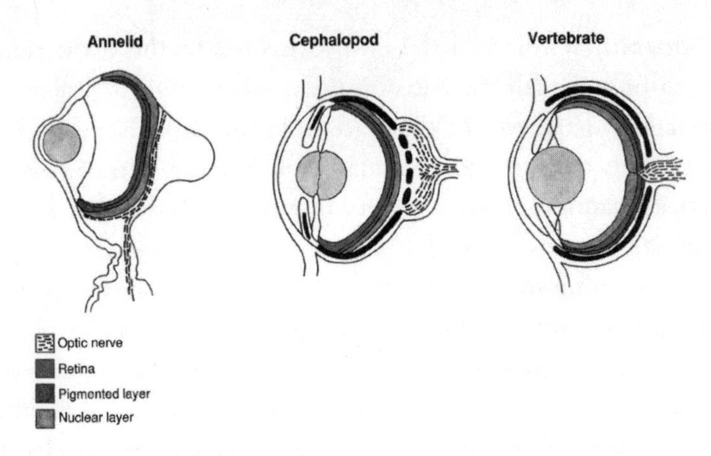

Figure 6.1: Convergent evolution of the eye. The cameralike eyes of worms, mollusks, and vertebrates look very similar. This is not, however, because they share common ancestry. The laws of optics have driven natural selection to find similar solutions to the problem of imaging in unrelated organisms. *(Professor Simon Conway-Morris. Taken from Simon Conway-Morris,* Life's Solution: Inevitable Humans in a Lonely Universe. *Reprinted with the permission of Cambridge University Press.)*

variety of biological problems. Cases in point are the cameralike eyes of vertebrates, cephalopod mollusks (such as the squid), and even some annelid worms (figure 6.1). All of these eyes consist of a light-limiting iris and a lens that focuses light on a retina packed with light-sensitive cells that relay information to the central nervous system. At first glance you might think it impossible that the exact same structure, the same organ, could have evolved twice, let alone three times. It hasn't, of course, because these three eyes are not identical in detail. The vertebrate eye develops as an outgrowth of the brain during the embryonic stage, which means that its retina is, in a sense, wired backward. The nerve connections of photoreceptors actually emerge on the side facing the light and must be drawn through a hole in the retina to reach the brain. The portion of the retina where this hole forms is a blind spot, and it limits the visual fields of all vertebrates. The mollusk eye forms differently, and its development allows the nerve connections to develop on the side of the retina away from the incoming light. It's wired the logical way, and it doesn't have a blind spot.

Once again, we might say that Gould was right. Several groups of organisms separated by more than 400 million years of evolution were faced with the problem of vision. The details of their evolutionary solutions were different—their light-sensing organs developed from different portions of the nervous system and were wired up in quite different ways. The proteins from which they formed lenses and transparent fluids were also different. But the physics of light and the laws of geometric optics were constraining factors in both cases and ultimately drove the process of natural selection. The final result, in each case, was so similar that we have to look closely into the details of structure and biochemistry to determine the ways in which these remarkable camera-eyes are actually distinct from one another. The convergence of form and function was nearly perfect. Score a big one for Conway-Morris.

My goal, however, is not to judge an intellectual competition between Gould and Conway-Morris, but rather to highlight the extent to which "random" evolutionary change operates within meiosis-like constraints to produce results that are unpredictable in detail but eventually become predictable in general outcome. This, I would argue, is the key to answering the objection that evolution is an unplanned project without meaning, purpose, or value. And the central question, of course, is not how closely Australian marsupials resemble mammals of the continental mainland, or what the key differences might be between our eyes and those of a squid. It is whether the appearance of humans, or something like them, was a certainty once the evolutionary process got going.

Gould clearly would be downright uncomfortable with such a notion, and once again, in detailed terms he would be absolutely right. From a biologist's point of view, the great tree of life spreads its branches in all directions, giving rise to simplicity as well as complexity, to the water flea as surely as to the philosopher. It would be an act of unbridled arrogance for us to examine the living history of this planet and pronounce ourselves, in Gould's words, "the summit of life's purpose." Run the tape of life again, starting from the Cambrian or wherever one might choose, and it's almost inconceivable that you'd get hairless bipedal primates with brains big enough to endow them with self-awareness, reflective thought, and calculus.

But, upon reflection, that's not really the issue.

The issue is not whether the exact scenarios of this planet's actual natural history would be repeated. They clearly would not. The genuine question is what sort of living world would emerge from a second or third running of the tape of life. Although we cannot predict the detailed outcome, this much we do know: Life vigorously explores adaptive space, and it finds its way to the same niches in that space again and again. One can, for example, study the ecological roles played by dinosaurs in various habitats—herbivores, scavengers, predators, keystone species—and discover that the exact same roles were quickly taken by mammals, birds, and reptiles when these great animals disappeared. Not even the most gifted naturalist could have looked at the world of the Cretaceous and predicted exactly how the balance of nature would settle in the postdinosaur world—but even the dullest would have been confident that settled it would be.

Turning our attention to the special case of our own species, we can be fairly confident, just as Gould tells us, that our peculiar natural history would not repeat, and that self-awareness would not emerge from the primates. Indeed, we would have no reason to suppose that primates, mammals, or even vertebrates would emerge in a second running of the tape. But as life reexplored adaptive space, could we be certain that our niche would not be occupied? I would argue that we could be almost certain that it would be—that eventually evolution would produce an intelligent, self-aware, reflective creature endowed with a nervous system large enough to solve the very same questions that we have, and capable of discovering the very process that produced it, the process of evolution. To argue otherwise would be to maintain, against all evidence, that our appearance on this planet was not the product of repeatable natural events. It would be to maintain, for no particular reason, that this corner of adaptive space was found once by the evolutionary process but could never be found again. Everything we know about evolution suggests that it would, sooner or later, get to that niche.

I'll admit that there's nothing to be gained by pretending that one can settle this question of repeatability with any certainty. So far as we know, nature has conducted the experiment just once, and the result was us (plus a few million other species). Science demands repeatability, and that's not possible in this case. Perhaps at some point in our own development we will discover a second experiment, a planet with characteristics similar to our own, on which we can truly test the grand principle of evolutionary convergence. Maybe that data will even be good enough to satisfy a Steve Gould. But that's a question for another book, and maybe even for another century. The point for today is that it's perfectly reasonable to maintain that evolution as we know and understand it was almost certain to produce a species like ours under conditions that prevail on Planet Earth.

MEANING

Rick Santorum's assertion that evolution tells us we are "a mistake of nature" makes me wonder whether Mr. Santorum is aware of the nature of physical law, as highlighted by the work of Martin Rees and many other scientists. It makes me wonder whether he has even considered the issue of evolutionary convergence. And it certainly makes me shake my head in amazement that someone living in the midst of the molecular revolution in biology could dismiss the creative power of evolution as nothing more than a series of mistakes.

Even the word "mistake" implies intentionality, albeit a failure of the power of intention. To say, for example, that a break in double-stranded DNA allowing two molecules to swap bases is a "mistake" is to imply that we know what the "intent" of DNA might be. That intent, presumably, is that such exchanges should never happen. Yet they happen all the time, and, in fact, there are no known genetic systems in which such swaps, or recombination events, do not take place. It would be just as fair, and maybe more accurate, to describe such events as "features" of the genetic system rather than as mistakes.

The language of chance and error is critical because we interpret these words in a way that causes us to disconnect from causality whenever we hear them. When we learn that something is a "random" event (whether a mugging or a lottery win), we immediately assume that whatever took place had nothing to do with the individual affected by the event. Therefore, if life itself was given us by evolutionary random chance, then we shouldn't bother searching for meaning in our own existence because we're not going to find any. We are just the product of random molecular and physical forces. We have no reason to regard our existence as anything but a pointless byproduct of nature—we are anything but "children of the universe."

As we have seen, this bleak view is actually at variance with what we know about the nature of our universe and the nature of evolutionary change. Life is possible only because of a precise balance of the fundamental constants that have built our universe. Whether this universe is one of many or the sole one that exists, it is the only one we know, and we are part of its fabric as surely as the spectrum of hydrogen or the valence of the carbon atom. As the distinguished astronomer George Coyne has said, we are literally made of stardust. The elements that enable life are formed in the stars themselves, and only a universe like ours could have forged these elements in abundance sufficient to produce planets where life was possible. That is indeed the universe in which we live, and it would be the height of scientific folly to maintain that such connections are without meaning or importance. Our connection with nature could not be more intimate, or more significant.

People of faith, of course, are fairly confident that they know what at least one aspect of that "significance" might be: namely, that ours is a universe willed by God, and that our presence within it is part of his plan and purpose. Religious understandings of nature and science play a critical role in our society, and I will discuss them shortly. It would be a serious mistake, however, to argue that the conditions that gave rise to our existence can be understood only in strictly religious terms. The great value of science is that it speaks to a common human culture of rational understanding, and that culture is accessible both within and outside of a religious context.

What I have argued for in this and in the preceding chapter is a kind of "evolutionary cosmology," a view of the universe in which our existence in the living world is an inherent part of nature itself. As a species we have a curious tendency to set ourselves apart from nature, and that tendency has led to some of our worst sins against the natural world. A clearer understanding of our relationship to nature leads, without question, to an appreciation of the unique role our species plays at this moment in the history of our planet. It is certainly correct to remind ourselves that we are just one species among millions that share the earth, but it is even more important to have a realistic view of our own value to that living world. We may be one among many, but we are the only one that can actually catalog and value that diversity, and the only one that can act in a way that will directly and consciously affect the future of this planet. That is an awesome responsibility, and it infuses every human life with meaning and value. If self-awareness is truly built into the evolutionary cosmology of existence, then the environmental crises we face today emerge, just as we do, from the stuff of stars and the vastness of the universe. If we are up to the task, if we deal wisely with challenges like global warming, extinction, and overpopulation, the test we will have passed will be a cosmic one in every respect.

Evolutionary cosmology not only promotes a powerful ethic of environmental stewardship, but it places an extraordinary premium on scientific learning and understanding. We are the only creatures who know where we came from; we are the ones who have split the atom, traveled to the moon, and unraveled genomes. In short, we have developed a view of the universe that is available to no other creature on earth. It is a fine thing to emphasize our deep kinship with all of life, but try to discuss quarks with your pet cat and you'll quickly appreciate what makes human existence so extraordinary.

Our species has opened a window on existence, called science, that we must never take for granted and never allow to be closed. Science, as done by human beings, is clearly dependent upon our unique biology, our highly developed nervous system, and our capacity for learning. But there is much more to science than our particular abilities as living organisms, for the conditions necessary for successful scientific inquiry transcend biology. They do not exist in all human societies and have been sadly absent for much of human history. A society that genuinely supports science is a rare and delicate thing. Science demands free inquiry, open discussion and debate, and popular support for the life of the mind. These are threatening principles to many human institutions and have been actively opposed by authoritarian regimes in the past and even in our own times.

Evolutionary cosmology places a special significance on the development of scientific culture, and a special responsibility on all of us to keep that culture going and to fight for the values that make science possible. If one were to search for meaning in a scientific framework, I can think of no higher calling than striving to ensure that human societies remain open and hospitable to the scientific way of thinking. Science is indeed a human activity, with all of the faults and failings that attend any human creation, but it is also a human creation that continues to promise to add greater insight, greater understanding, and greater depth to our lives.

There is an additional element of evolutionary cosmology that should not be underestimated or devalued, although it often is—its ability to give pleasure. There are few pursuits in life more exhilarating than the search for knowledge, and few rewards more gratifying than its attainment. To investigate and to solve a question, no matter how small, gives one an extraordinary sense of satisfaction. This is something that all experimental scientists understand, although far too few can communicate it emotionally to those not directly involved in science. Nonetheless, science is a collective process. Individuals make discoveries and advances, but they do so within the framework of ideas, tools, and concepts developed by others. This means that the pleasures of understanding, ultimately, are available to everyone who cares to learn, think, and follow our advancing understanding of the natural world.

If there is one value that should be common to all human cultures, it is that knowledge is to be preferred to ignorance. Such knowledge can come in many forms. Although in the popular imagination scientific knowledge is cold, rational, objective, and dispassionate, this view is dramatically contradicted by Einstein's remarkable statement that "the most beautiful thing we can experience is the mysterious—it is the source of all true art and science." The apparently disparate worlds of art and science are drawn to the mysterious because in different ways they both explore the unknown. Both disciplines, however, ultimately center on human experience, and they both seek to crystallize that experience into understanding. To experience the level of understanding produced by the greatest art and the most profound science is one of the deepest joys of life, and surely among its most significant.

Finally, what are we to make of the "purpose" of life? Some of my scientific colleagues, especially those who reject a theistic view of our existence, have argued that the question of life's purpose is not even worth asking. I have been told that the question of meaning is itself without meaning (whatever that might mean), and that we humans ask such questions only because our evolutionary heritage programs us to look for a hidden significance to events, perhaps as a survival trait.

Whenever I hear this objection I am reminded of Ray Bradbury's marvelous science fiction tale *The Martian Chronicles.* In Bradbury's story our neighboring planet has been successfully explored and colonized, and humans have quickly subdued and dominated the planet. Native Martians are gradually forced away from our settlements, and eventually the earthlings come to believe that Martians have become extinct. In the novel's climax we humans revert to our customary ways of bloody conflict, and suddenly Mars is left uninhabited—except by the long-hidden Martians, who now return to reclaim their planet. As they return, among the earthling traits the Martians find most puzzling is the human obsession with the purpose of life, and they give thanks that they do not share this puzzlement. The purpose of life, as every Martian instinctively realizes, is to be alive.

It would be hard to improve on the insights of Bradbury's Martians. The gift of life itself supplies a purpose so compelling and profound that all else follows from it. The lesson of evolutionary cosmology is that each moment of our existence is a gift from the stars, to be experienced, treasured, used wisely, and enjoyed. That is how today I would answer my daughter's question of what people are for.

GOD

When I served as the opening witness in the September 2005 trial in federal court in Pennsylvania on the issue of teaching "intelligent design" in public schools, I was critical of the notion that ID is authentic science, and I opposed the attempts of political authorities (the local school board) to force teachers to insert it into their lessons. The trial was highly publicized, and naturally it provoked strong reactions on both sides of the issue. Over the next few days plenty of people who took issue with my two days of testimony let me know exactly how they felt, in letters, in e-mails, and even by telephone. Hostile reaction is to be expected whenever one takes a public stand on a controversial issue, and quite frankly, I'm used to it. What struck me about the reaction to Dover, however, was the religious character of almost every hostile comment. In addition to being told where I would likely spend eternity (no need for warm clothes, one e-mail assured me), I was repeatedly lectured about the disrespect that I and other scientists had shown for the Almighty. Evolution, according to these critics, takes God out of the picture, and therefore must be opposed by people of faith at all costs. How dare I call myself a Christian and speak on behalf of Darwin?

It is not my purpose to deal in depth here with the theological issues that relate to evolution and modern science. Readers who might be interested in such questions may look at my earlier book *(Finding Darwin's God: A Scientist's Search for Common Ground Between God and Evolution)* or consult the thoughtful analyses of theologian John Haught (whose works include *God After Darwin* and *Responses to 101 Questions About God and Evolution*).[11] Nonetheless, the conviction that evolution is anti-God is so widespread that at least a few words of comment are required.

Some of this hostility has surely been generated by those who choose to read the creation accounts of Genesis literally. If one does indeed take the position that the universe is a mere six thousand years old, then one's conflict is not just with evolution, but with all of modern science. As I have explained elsewhere,[12] the physical and biological evidence for today's view of natural history is overwhelming. The earth is billions of years old, and the universe several times older than that. Life did not appear on this planet in a single burst of creative energy, but it appeared gradually, step by step, over hundreds of millions of years. In so doing it left behind an extraordinary record of ancestor-descendant relationships that tell an evolutionary story of our own development and that of every other living thing. In short, a literal reading of the Genesis story is simply not scientifically valid.

Genesis was written in a prescientific age, in the language of the day and in an attempt to communicate great truths to the people of that age. Those truths include above all the notion that we are here along with all other existence as the result of the creative power of God. They do not include an attempt to teach science. This was exactly the point made decades ago by the late Pope John Paul II:

> The Bible itself speaks to us of the origin of the universe and its make-up, not in order to provide us with a scientific treatise, but in order to state the correct relationships

> of man with God and with the universe. Sacred Scripture wishes simply to declare that the world was created by God, and in order to teach this truth it expresses itself in the terms of the cosmology in use at the time of the writer. The Sacred Book likewise wishes to tell men that the world was not created as the seat of the gods, as was taught by other cosmogonies and cosmologies, but was rather created for the service of man and the glory of God. Any other teaching about the origin and make-up of the universe is alien to the intentions of the Bible, which does not wish to teach how heaven was made but how one goes to heaven.[13]

John Paul II's view of Genesis and the nature of creation is a completely traditional one within the Christian context. St. Augustine, one of the most influential and prolific of the early Christian writers, argued essentially the same point. In his remarkable fifth-century book *On the Literal Meaning of Genesis,* Augustine specifically warned Christians against using Scripture to make statements about astronomy, biology, and geology. The worst thing that could happen, he advised, would be for nonbelievers to hear Christians, "presumably giving the meaning of Holy Scripture, talking nonsense on these [scientific] topics."[14] Augustine urged his readers to do everything they could to prevent well-meaning Christians from casting disrepute on Scripture by reading the Bible as a scientific document.

Augustine, of course, was not an advocate of evolution, having lived fourteen centuries too soon to know the work of Charles Darwin. But Augustine was familiar with the science of his own day and clearly recognized the dynamic nature of the universe. He understood, in simple terms, that creation was not a finished project, but a continuing one:

> The universe was brought into being in a less than fully formed state, but was gifted with the capacity to trans-

> form itself from unformed matter into a truly marvelous array of structure and life forms.[15]

Augustine's insight is critical today, because his recognition of the self-transforming character of nature speaks to the heart of the unease that so many Christians feel about evolution. As one of my letter-writing critics charged, it is the "height of scientific arrogance" for people like me to maintain "that nature itself can create life." And yet Augustine says exactly the same thing, and finds it very much in the Christian tradition. How can this be? Would it be too much to suggest that early Christian thinkers actually had a more expansive view of the relationship between God and his creation than many Christians do today?

On the day of his coronation as pope, Benedict XVI (formerly Joseph Ratzinger) made a statement that was widely interpreted as supporting the intelligent design movement against evolution. The new pope assured his audience that "we are not some casual and meaningless product of evolution. Each of us is the result of a thought of God. Each of us is willed, each of us is loved, each of us is necessary."[16]

While I will not venture a guess as to what Pope Benedict's ultimate record will be with respect to science, his comment was clearly not the beginning of an antievolution trend within the Catholic Church, as some had hoped. Science assures us that we are indeed the products of evolution, but it certainly does not tell us that we are its "casual and meaningless" products. The scientific insight that our very existence, through evolution, requires a universe of the very size, scale, and age that we see around us implies that the universe, in a certain sense, had us in mind from the very beginning. It is very much, in the memorable words of physicist Freeman Dyson, a "universe that knew we were coming."[17]

If this universe was indeed primed for human life, then it is only fair to say, from a theist's point of view, that each of us is "the result of a thought of God," despite the existence of the natural processes that gave rise to us. The skeptic will object to this viewpoint, of course, regarding it as nonscientific. And so it is. Theology does not and cannot pretend to be scientific, but it can require of itself that it be consistent with science and conversant with it. Remarkably a theology consistent with evolutionary cosmology is more possible today than ever before, and we do indeed have science to thank for that.

When I am asked how I manage to "reconcile" evolution with religious faith, I often shake my head and try to explain that I don't "reconcile" them at all. If two ideas are not in conflict, they have no need of reconciliation. But isn't there a basic conflict with faith in the notion that God needed natural processes to accomplish his ends? Here again I would turn to Augustine. Questioned about the role that natural processes might play in accomplishing the will of God, he wrote:

> We admit that what is contrary to the ordinary course of human experience is commonly spoken of as contrary to nature.... But, God the Author and Creator of all natures does nothing contrary to nature; for what is done by Him who appoints all natural order and measure and proportion must be natural in every case.[18]

As Thomas Aquinas was later to point out, if God exists, he is the author of nature itself, and the cause of causes. Therefore, finding a natural cause for any phenomenon does not take it out of the realm of divine providence. If all of nature is part of God's providential plan, as the Western monotheistic religions teach, then the science of natural cause exists within that providence and does not contradict it.

The great concern of scientific materialists is that the tenets of religion will weaken and undermine empirical science. They cite the historical persecutions of scientific figures like Galileo and the present-day hostility of large segments of the religious community to science as evidence that their fears are well founded. The great concern for people of faith is that science is deeply hostile to religion in any form, and that scientific materialism's real purpose is the destruction of religious faith and the demoralization of society. They cite as evidence the lack of religious faith among elite scientists and the string of books written by leading intellectuals that employ scientific concepts as weapons against religion.[19] As is often true in such conflicts, extremists on both sides make statements that justify opponents' worst fears. The great irony of our present situation is that it has never been easier to see why this conflict is unnecessary.

The evolutionary cosmology that emerges from physics and biology tells us that we are indeed made, just as Scripture claimed, from the dust of the earth itself. But the details of that story are grander than any of the authors of Scripture might have dreamed. For human life to have developed on our planet, we need a universe even vaster than the nighttime sky. We require a cosmos of inconceivable age, finely tuned fundamental constants to stoke the fires of trillions of suns, and a balance of light and heavy elements forged in the embers of dying stars. And we do indeed have all of them.

These ingredients, in the words of Martin Rees, make up a "recipe" for our universe, a step-by-step guide to building an existence in which the capacity for life emerges from the very forces that shape galaxies and fill the earth with light and color. Where did this fortunate collection of circumstances come from? To the faithful, they came, as all things do, from God.

One might conclude that the hypothesis of God puts the believer and nonbeliever at odds with respect to evolutionary cosmology, but this is not the case. Science draws its meaning and value from the search for truth about the natural world, and in this context it has told us, at least so far, that we are every bit the "children" of the universe that "Desiderata" assured us we were. Believers and nonbelievers can agree on this element of science, and then part company as to how it is to be interpreted. Yet, on this important point they still have plenty to say to one another. Nonbelievers should remind the religious that, however they might take comfort in the words of Jesus or Aquinas or Benedict, their faith is not scientific, nor can it become scientific. Therefore they should not use their faith to bend the conduct or findings of science in any way. But believers have something to say, as well, and they are right to remind skeptics and agnostics that one of their favored explanations for the nature of our existence involves an element of the imagination as wild as any tale in a sacred book: namely, the existence of countless parallel simultaneous universes with which we can never communicate and whose existence we cannot even test. Such belief also requires an extraordinary level of "faith," and the nonreligious would do well to admit as much.

Whether one embraces the teachings of religious faith or not, the claim that our universe "knew we were coming" stands as a perfectly valid metaphor for the deep linkages between our personal existence and the nature of reality. If we occupy today a special place in the universe, as I believe we do, that place is, in the words of St. James, "that we should be a kind of first fruits of his creatures."[20] We are the first organisms in our neighborhood of the cosmos to gain a partial understanding of the fabric of nature, and we should make the most of it.

SEVEN

Closing the American Scientific Mind

AMERICA IS the greatest scientific nation the world has seen, and there is a reason why this is so. Science has thrived in the United States because the scientific way of thinking is part of the American character—practical, pragmatic, and based on what you are able to do, not who you are. Success in America has been based not upon appeals to nobility, lineage, or authority but on direct, tangible results. One of my European friends once confided in me that the frightful thing about America was that each generation was expected to prove itself, time and time again, and always under different circumstances. Frightful indeed—but remarkably like the practice of science, in which the value of your latest work or discovery is the only currency that can buy prestige and advancement. One of my undergraduate professors had a sign posted in his lab stating, NOTHING IS MORE TERRIBLE THAN THE MURDER OF A BEAUTIFUL THEORY BY A GANG OF UGLY FACTS. True indeed—but scientists are always, always on the side of those thugs, no matter how many beautiful theories they are inclined to beat down.

When I meet with my teaching assistants in the large introductory biology course I teach each spring at Brown, I confess to them that I take an evangelical approach to the teaching of biology.

The word "evangelical" shocks them, because in modern America they've learned to associate it with religious fundamentalism, but they shouldn't. The word actually means nothing more than speaking the truth and bringing that "good message" of truth to others—which is exactly how I ask my assistants to approach the course. One year I remember telling my staff something like this: "I teach at a university, so I do know there are other disciplines. I also know that some of our students will major in history or art or chemistry. But for the life of me, I cannot understand why any young person at this point in time would want to study anything other than biology—and that's exactly the attitude I want you to convey to our students."

I hope you won't get the wrong impression from this confession. I do indeed appreciate the value of the humanities and social sciences, as well as the other natural sciences, and I feel fortunate to teach at a university where the scale of our campus is small enough to actually allow me to know scores of colleagues in these fields. I expect them to present their own disciplines to students with the same evangelical fervor that I try to apply to the science of life. Learning is nothing without passion, and I would hope that every person privileged to instruct college-age students would pass along the same passion that led them to select their chosen field.

Like any person who has been lucky enough to pursue a career in science, I've had a chance to experience the ups and downs of research, the joys of discovery, and the disappointments of failure. I am quick to tell relatives and friends that I have the best job in the world, one that grants me both the responsibilities of teaching some of the best students in the world and the incredible opportunities of walking into the lab each morning and wondering, "What should I try to discover today?"

One of the elements of science that I have always valued is its democratic nature, the fact that literally anyone can make a contribution to research, and that sooner or later that person's contribution will be judged on its merits. This is not to say that the scientific enterprise operates without racism or sexism or class and personal prejudices—but it is to say that the ultimate judge of authority in science is nature itself. In the final analysis bad science fails, and good science wins on the basis of the evidence, on the natural authority of observation and experiment. As a practical matter this has made science one of the few areas of academic inquiry that is largely unaffected by politics, whether in the so-called real world or in academia. Once again that is not to say that science does not develop a personal politics of its own, like any social activity, or that scientists are politically neutral—they certainly are not. But it is to say that the practice of science, when compared to that of other academic disciplines, has been remarkably free of politics and political considerations. The answers I sought in my research, just like those of any other investigator, were the right answers, not liberal ones or conservative ones or progressive ones. There is no such thing as a socialist view of the Krebs cycle, or a capitalist explanation for DNA replication, nor should there be. Science isn't pure, but science and mathematics are nonetheless the closest things we have on this planet to a universal culture or, perhaps more accurately, to a search for knowledge that transcends culture.

For most of my career this is exactly the position that science has held in academic life. There have been a few intrusions, to be sure, and early on I witnessed a high-profile political battle take place in the very institution where I worked. I began my faculty career at Harvard in the 1970s, just as the eminent biologist E. O. Wilson published a groundbreaking book called *Sociobiology,* which suggested that human social behavior was directly accessible to biological analysis. In the Harvard of the 1970s such ideas were seen as having distinctly political connotations, and Wilson was vigorously attacked by faculty associated with the academic left, which included fellow evolutionary biologists Richard Lewontin and Stephen Jay Gould. Because my own small lab was across the street from these luminaries (in a department of cellular and developmental biology), I never directly observed any of the widely rumored hallway confrontations that animated this debate. Despite this it was impossible not to notice just how heated the rhetoric became, or how deeply politics had begun to disrupt the desires of Wilson and many others simply to do their science regardless of outside pressures.

Eventually the "sociobiology wars" of the 1970s ended, and when I moved to Brown a few years later, I fully expected that the political squabbles I had seen in Cambridge had been a curious exception to the general way in which science is practiced. And that was almost true, at least for a while. There is no question, of course, that issues of how the *findings* of science should be applied practically through technology have always had political implications. Atomic power and nuclear weapons are clear examples of this, as are the current ethical debates regarding embryonic stem cell research. However, the arguments do not directly take issue with the underlying science. Those opposed to nuclear weapons do not dispute the existence of the weak nuclear force or the equivalence of matter and energy. Those who would ban the use of embryonic stem cells do not claim that such cells don't exist, but rather that the techniques used to obtain them are not ethical. Politics in these cases does not dispute the authority of science itself, which means that it does not pose a direct challenge to the process of scientific inquiry.

If anyone could have taken comfort in this analysis, however naïve it might seem, that confidence vanished with the appearance of the intelligent design movement. When I first encountered the term, I saw that most of its arguments were little different from the creation science arguments that had preceded it in the 1970s and 1980s. Before long, however, it became apparent that there was something truly different about ID. Its goals went far beyond merely carving out a little scientific respectability for antievolutionism, and extended all the way to a complete redefinition of the way that science—all of science—is done. The proponents of ID seek nothing less than a true scientific revolution, an uprising of the first order that would do a great deal more than just displace Darwin from our textbooks and curricula. They seek the undoing of four centuries of Western science, and that surely should be enough to make anyone sit up and pay attention.

OPENNESS—THE AMERICAN VIRTUE

I had been teaching at American universities for more than a decade when in 1987 a remarkable book directed at the heart of university life took the best-seller lists by storm. It had been written by Allan Bloom, professor of philosophy at the University of Chicago, and it bore the extraordinary title *The Closing of the American Mind.* Bloom's subtitle left no doubt as to the target of his book: *How Higher Education Has Impoverished America's Young and Failed Its Students.* As a young college professor who didn't always like what he saw within the halls of ivy, the book caught my eye immediately.

Bloom's work was widely, and correctly, interpreted as a frontal assault on the culture of the academic left in American colleges and universities. As such it was generously praised by political conservatives, especially those outside academia. They had never much liked the changes in higher education associated with progressive movements like feminism and affirmative action during the 1960s and 1970s, and Bloom's thesis that there was something deeply wrong in the university was sweet confirmation of their discontent. Predictably those at the forefront of introducing those changes reacted differently. They accused Bloom of playing reactionary politics with the hard-won achievements of these decades, and saw his book as an effort to reverse the gains that women and minorities had made in academic life. The themes of his analysis are still relevant to American higher education, and his book remains widely read even today.

It is not my purpose to reanalyze Bloom's critique of American education, except for the brilliant metaphor of the book's title—the *closing* of the American mind. What could he have meant by that? Weren't the changes of the 1960s and 1970s, regardless of what one thinks of them, all in the direction of openness? By any standard isn't higher education today much more open to new ideas and cultures than it was in the 1940s and 1950s? The answer, of course, is yes, whether one thinks ill or well of those developments. So, how could Bloom possibly have characterized the process as a "closing" of the mind? This, of course, was the core of his thesis: The American academic mind had become closed precisely because it had become too open.

The first chapter in Bloom's book is entitled "Our Virtue," by which he means the characteristic American virtue of openness. Yes, he acknowledges, American minds are open, and they are particularly open at the university, but not in the way one might hope. Bloom believes that that is the case because American students have been taught, under the new academic regime, to fear not error but something else entirely:

> The danger they have been taught to fear is not error but intolerance. Relativism is necessary to openness, and this is the virtue, the only virtue, which all primary education for more than fifty years has dedicated itself to incubating. Openness—and the relativism that makes it the only plausible stance in the face of various claims to truth and various ways of life and kinds of human beings—is the great insight of our times.... The point is not to correct the mistakes and really be right, rather it is not to think you are right at all.[1]

Under this set of academic values, the worst sort of bigotry is associated with coming to a decision on the basis of evidence. To do so would be to reveal that one is not "open" to other ideas, and that would be an academic sin of the first order. The difficulty with this new academic value is clear to Bloom:

> Openness used to be the virtue that permitted us to seek the good by using reason. It now means accepting everything and denying reason's power. The unrestrained and thoughtless pursuit of openness, without recognizing the inherent political, social, or cultural problem of openness as the goal of nature, has rendered openness meaningless.[2]

Properly understood, Bloom's analysis does not fall into either of the neat categories of "liberal" or "conservative." The danger of the academic left is not that it is left-wing, but rather that it has denied the power of reason to seek truth and the good life. The results of this quest for an uncritical openness to all things have been tragic, according to Bloom, because they have deprived Western intellectual life of its most important virtue:

> What is most characteristic of the West is science, particularly understood as the quest to know nature and the consequent denigration of convention—i.e., culture or the West understood as a culture—in favor of what is accessible to all men as men through their common and distinctive faculty, reason. Science's latest attempts to grasp the human situation—cultural relativism, historicism, the fact-value distinction—are the suicide of science. Culture, hence closedness, reigns supreme. Openness to closedness is what we teach.[3]

Bloom uses the word "science" here in its broadest sense, meaning the systematic search for truthful answers to questions in all fields of inquiry. (He reserves the term "natural science" for the likes of biology, chemistry, and physics.) He particularly disparages the latest fruits of this endeavor, most especially the cultural relativism that now pervades many disciplines. As a result, individual expression has itself become devalued to the point of meaninglessness:

> The rights of science are now not distinguishable from the rights of thought in general, or any description whatsoever. Freedom of speech has given way to freedom of expression, in which the obscene gesture enjoys the same protected status as demonstrative discourse. It is all wonderful; everything has become free, and no invidious distinctions need to be made.[4]

To Bloom the great tragedy of modern intellectual life is that the "openness" found in America's universities is actually an unwillingness to apply reason to solve human problems. Not to choose among alternatives is to accept that all ideas, all ethical systems, all historical constructs are equally valid. And if that is true, why bother to make such intellectual constructs at all? If every individual can build his own system of values and principles without reference to the great ideas of the past, then why even bother to grapple with the likes of Socrates and Plato and their pesky insistence on reasoned dialogue? The paradox, of course, is that such openness to all ideas has closed the American mind to the possibility of using higher education to choose, of honing the faculty of critical reasoning to achieve genuine knowledge of any question worth answering. And a mind that refuses to do that is indeed closed, in the most tragic sense. Our virtue has become our greatest vice.

At the time that Bloom wrote, he believed that the politics of openness had affected just about every part of the university—except for the sciences.

> How are they today, the big three that rule the academic roost and determine what is knowledge? Natural science is doing just fine. Living alone, but happily, running along like a well-wound clock, successful and useful as ever. There have been great things lately, physicists with their black holes and biologists with their genetic code. Its objects and methods are agreed upon. It offers exciting lives to persons of very high intelligence and provides immeasurable benefits to mankind at large. Our way of life is utterly dependent on the natural scientists, and they have more than fulfilled their every promise.[5]

Why had the natural sciences escaped the intellectual plague that Bloom found in the humanities and social sciences? The key, as Bloom saw it, was in their uncompromising devotion to empirical values, to their use of nature as the ultimate standard of proof.

> Natural science simply does not care. There is no hostility (unless it is attacked) to anything that is going on elsewhere. It is really self-sufficient, or almost so. If some other discipline proved itself, satisfied natural science's standards of rigor and proof, it would be automatically admitted. Natural science does not boast, it is not snobbish. It is genuine.[6]

From Bloom's point of view science was immune to the sort of attacks that had undermined the humanities, the claims that alternate cultures and viewpoints had been unfairly excluded from the canon of learning. There could be no persuasive value in applying such arguments to the sciences, because the only thing that mattered in these disciplines was the nature of the evidence. As Bloom put it, "The natural sciences are able to assert that they are pursuing the important truth, and the humanities are not able to make any such assertion. That is always the critical point."[7]

But Bloom was wrong. What he could not appreciate, back in 1987, was that a new attack on the primacy of science was incubating, and in a location intimately familiar with the very tactics and strategies that had led to the situation he decried in the humanities. The place was Berkeley, California, and the person leading it was a tenured professor at the University of California. His name was Phillip Johnson, and his specialty was criminal law.

To Johnson, for whom evolution had been a stumbling block on the road to faith, Darwinism represented much more than just an erroneous scientific theory. It was the essential underpinning of a worldview that was profoundly hostile to religious thought, and therefore to the founding principles of Western society. Johnson found a willing and well-funded sponsor for his ideas in the Discovery Institute of Seattle. Inspired and guided by Johnson, Discovery established a Committee for the Renewal of Science and Culture to provide intellectual and financial support for the leading advocates of ID. Quite consciously Johnson then set about developing a strategy that would subject science to the same relativistic critique that had already affected the rest of academia. That strategy would become known as the Wedge, and its great irony was that it borrowed the language and tactical brilliance of the academic left—except that this time these rhetorical weapons would be wielded on behalf of the academic right. Turnabout is, after all, fair play. And "fairness" was at the heart of Johnson's new idea.

PLAYBOOK

In 1998, in an effort to attract funding and support, Johnson assembled his ideas in an outline known as the Wedge document, shortly after a highly publicized debate on evolution aired on *Firing Line,* William F. Buckley's show on the Public Broadcasting Service (PBS). The Wedge document became public when it was leaked to the Internet in 1999, although several years would pass before the Discovery Institute grudgingly admitted its authenticity.[8]

The history of this document has been thoroughly researched by Barbara Forrest, and it is detailed in her book, coauthored with Paul Gross, *Intelligent Design: Creationism's Trojan Horse.* The first two paragraphs of "The Wedge" state its core assumptions clearly:

> The proposition that human beings are created in the image of God is one of the bedrock principles on which Western civilization was built. Its influence can be detected in most, if not all, of the West's greatest achievements, including representative democracy, human rights, free enterprise, and progress in the arts and sciences.
>
> Yet a little over a century ago, this cardinal idea came under wholesale attack by intellectuals drawing on the discoveries of modern science. Debunking the traditional conceptions of both God and man, thinkers such as Charles Darwin, Karl Marx, and Sigmund Freud portrayed humans not as moral and spiritual beings, but as

> animals or machines who inhabited a universe ruled by purely impersonal forces and whose behavior and very thoughts were dictated by the unbending forces of biology, chemistry, and environment. This materialistic conception of reality eventually infected virtually every area of our culture, from politics and economics to literature and art.[9]

So, what was to be done in response to this secularization? "The Wedge" outlines a series of strategic moves designed to split science apart from its links to "materialism" and "to replace it with a science consonant with Christian and theistic convictions." One might argue, as I have, that science is already consonant with Christian and theistic convictions, but the advocates of intelligent design disagree. To them today's science is bitterly hostile to religion and needs a fundamental overhaul. The Wedge strategy proposes a series of projects, including efforts in research, writing, public relations, and "cultural confrontation," to achieve that aim.

> If we view the predominant materialistic science as a giant tree, our strategy is intended to function as a "wedge" that, while relatively small, can split the trunk when applied at its weakest points. The very beginning of this strategy, the "thin edge of the wedge," was Phillip Johnson's critique of Darwinism begun in 1991 in *Darwinism on Trial,* and continued in *Reason in the Balance* and *Defeating Darwinism by Opening Minds.* Michael Behe's highly successful *Darwin's Black Box* followed Johnson's work. We are building on this momentum, broadening the wedge with a positive scientific alternative to materialistic scientific theories, which has come to be called the theory of intelligent design (ID).[10]

Unlike the earlier efforts of creation science, the Wedge playbook does not plan to stop once it has defeated "Darwinism," and in this respect the goals of the intelligent design movement are far broader than those of its creationist forebears. The real target of the movement, as the document makes clear, is the whole of science and the materialist foundations of the scientific process.

Had Bloom been around to read "The Wedge," he would have understood it instantly. In his view the academic left had not taken over the humanities by trying to reform them, by targeting one or two specific fields, or by trying to correct a little discrimination here and a little narrow-mindedness there. It had won its victories in the 1960s and 1970s by claiming that the whole intellectual enterprise was rotten to the core, and that the day of a new paradigm, a new way of thinking, had arrived. The first step along that road was to challenge the claim that traditional academic scholarship was a search for the truth. Once that had been accomplished, once students and their professors had become afraid of making up their minds about nearly anything, it was possible to introduce the new relativism that treated all cultures and all ideas as equally worthy. The Wedge strategy is nothing less than a playbook for doing exactly the same thing to the sciences.

GAME DAY

When I was a high school student, a rumor spread that a student who had transferred to another school had taken one of our football team's playbooks with him and had provided it to his new coach. Among our players, many of whom were my friends, there was anger and apprehension at the thought that a competing team might be privy to our signals and formations and plans. As the anxiety spread, our own coach called a team meeting to calm down his players. He was sure that the rumor wasn't true, but even if it was, he wasn't worried. Playbooks aren't nearly as important as how you perform on game day, he told his team. Watch us under pressure, he assured them, and you'll see something that isn't in any playbook.

That was good advice, and it applies equally to the intelligent design playbook. In the game that followed, its proposals weren't followed exactly. It may at one time have indicated the plans and intentions of the movement, but plans have a way of being constrained by reality, as any team's coach would admit. Some game plans work, and some don't. But the way in which the Wedge strategy has been played out provides a particularly dramatic demonstration of where the ID "team" is powerful and strong, and where it is demonstrably weak.

On one level, the scientific team put on the field to deploy the Wedge strategy has failed miserably. The playbook included short-term and long-term goals in a variety of areas, one of which was "scientific research, writing, and publicity," and among the specific five-year objectives was the publication of "one hundred scientific, academic, and technical articles by our fellows." The fellows of the Discovery Institute have indeed been busy writing, but none of them has produced a single scientific research paper that supports intelligent design. While they've written plenty of opinion pieces, reviews, op-eds, and position papers, their goal of producing a significant amount of scientific work has apparently been abandoned. It's almost as though a coach bragged about how he'd control the football by using running plays, but once the game was under way, he called for nothing but passes.

You don't have to take my word for this, or the word of any opposing scientist. As we have already seen, Phillip Johnson, interviewed in 2006, was remarkably candid about the failings of this aspect of the Wedge game plan. Eight years into the Wedge strategy (and three years after the five-year plan was to have been completed), he lamented that the movement's "scientific people" had yet to produce a legitimate scientific "alternative" that could stand on the same field as evolution. Johnson admitted, "No [ID] product is ready for competition in the educational world."[11]

Johnson's frank admission that his "scientific people" have not been able to produce a genuine alternative to evolution has been echoed by others in the ID community. Paul Nelson, a senior fellow of the Discovery Institute, conceded a more specific point in a 2004 interview with *Touchstone* magazine:

> Easily the biggest challenge facing the ID community is to develop a full-fledged theory of biological design. We don't have such a theory right now, and that's a real problem. Without a theory, it's very hard to know where to direct your research focus. Right now, we've got a bag of powerful intuitions, and a handful of notions such as "irreducible complexity" and "specified complexity"—but, as yet, no general theory of biological design.[12]

This lack of a "general theory of biological design" was especially evident at the Dover ID trial. Called as an expert witness to defend the policy of introducing students to ID, Michael Behe (also a Discovery Institute fellow) drew a blank when he was asked to tell the court about scientific successes in the ID research program. Federal Judge John E. Jones III, summarizing Behe's testimony, wrote:

> On cross-examination, Professor Behe admitted that: "There are no peer-reviewed articles by anyone advocating for intelligent design supported by pertinent experiments or calculations which provide detailed rigorous accounts of how intelligent design of any biological system occurred." [22:22–23 (Behe).] Additionally, Professor Behe conceded that there are no peer-reviewed papers supporting his claims that complex molecular systems, like the bacterial flagellum, the blood-clotting cascade, and the immune system, were intelligently designed. [21:61–62 (complex molecular systems), 23:4–5 (immune system), and 22:124–25 (blood-clotting cascade) (Behe).] In that regard, there are no peer-reviewed

> articles supporting Professor Behe's argument that certain complex molecular structures are "irreducibly complex." [21:62, 22:124–25 (Behe).] In addition to failing to produce papers in peer-reviewed journals, ID also features no scientific research or testing.[13]

If the forces executing the Wedge strategy fell about a hundred papers behind the scientific objectives of their five-year plan, other aspects of their program played out even better than they might have hoped. They expected "significant coverage" in national media, which they have certainly received, as ID became a regular topic on radio and TV talk shows, and even broke through for a cover story in August 2005 when *Time* magazine ran a feature called "The Evolution Wars."[14]

They also met another key five-year goal in their wish to see "major new debates in education, life issues, legal and personal responsibility pushed to the front of the national agenda." Indeed they were. Arguments based on ID surfaced in scores of school districts around the country, and pressure mounted on local boards to introduce ID in the science classroom as an alternative to Darwinism. Forces favorable to ID took control of the Kansas Board of Education in 2005 and in Ohio succeeded in placing ID-favorable language into state-approved lesson plans. Pennsylvania Senator Rick Santorum even managed to get ID-friendly language inserted into the Senate version of a bill that implemented President Bush's groundbreaking efforts at education reform. Although that language was deleted before the bill became law, parts of it remained in the conference committee report that reconciled Senate and House versions of the bill. This allowed ID proponents to claim, persuasively but inaccurately, that the new education law required the examination of scientific alternatives to evolution such as intelligent design.

I could list a series of other ways in which the five-year plan of "The Wedge" has met with success, but all fall into a single aspect of their strategy: namely, public relations. ID has been moved brilliantly into the center of the culture wars, and most polls show that a majority of Americans now support the teaching of ID in schools along with evolution. This is, significantly, one way in which the ID movement is radically different from the academic movements of the 1960s and 1970s. The advocates of intelligent design are not really interested in the academy. Their targets are education in grades K–12, where public opinion (which controls local school boards) really matters. While their scientific game remains nonexistent, their public relations efforts are grinding out yard after yard, and show no signs of being stopped.

NO REFUGE

Allan Bloom believed that the natural sciences would be immune to attack by virtue of the fact that they are "genuine," relating directly to the natural world and pursuing what everyone regards as "important truth." I suspect that, because of this perspective, he could not have foreseen the way in which the argument of relativism might be used to undermine them as well. In the case of evolution the first steps were calls for "balance" and "openness" in the science curriculum, particularly in the portion under popular control in the public schools. Louisiana's "creation science" advocates made this point clear in 1981 when they managed to persuade their state legislature to enact a "Balanced Treatment" act to promote their views.[15] If evolution is just one explanation for the origin of species, there surely must be others that could "balance" it, and why shouldn't schoolchildren be exposed to them? When it became clear that the alternate "explanation" of creation science was inherently religious, the Supreme Court struck down the Louisiana law. But the Court's clear decision in this case didn't diminish the appeal of the argument for fairness and balance. When the ID strategy was devised in the late 1980s, its enthusiasts saw no reason not to employ the tactic once again.

Once the scholars of ID had published a series of books and

essays stating their arguments in print, it became a simple matter for opponents of evolution to call for education to be "open" to these "new theories" and the "controversies" swirling around them. Never mind that these ideas had gained no credence within science and that therefore science did not regard evolution as controversial in any sense. But as Bloom had noted with respect to the opening of the academy, Wedge advocates argued that the greatest error was not to be mistaken, but to be close-minded to alternative ideas and interpretations. If science is willing to investigate all possible explanations, their argument went, it certainly should be open to the ideas of intelligent design. Not to allow these persuasive arguments into curriculum and classroom would be to enforce a stifling dogma on science teaching. The "dogma" of evolution was thereby attacked just as effectively as had been the primacy of Western literature, philosophy, and culture.

The burden then fell upon the scientific and educational establishments to explain why they were unwilling to consider these challenges. Not surprisingly they were just as hard-pressed to make the case for exclusion as university academics had been years earlier to make the case for the relative value of Western culture and philosophy.

But this was only the beginning. "The Wedge" had powerful, if unwitting, intellectual allies in its fight against evolution who were already present in the academy. These came from a fashionable school of scientific criticism wrapped in the idea of "cognitive relativism." Cognitive relativism is essentially an argument that truth and falsehood are not the absolutes we often take them to be, but rather are constructed by individuals and social groups. Karl Popper, the great philosopher of science, had helped to pave the way for this relativistic foray into science by challenging its cherished notion of objectivity: "The choice between competing theories is arbitrary, since there is no such thing as objective truth."[16] If objective truth is a myth, then what does this make of science in general? The answer should be perfectly clear: not much.

Cognitive relativism was applied to science with special vigor by Paul Feyerabend in his book *Against Method*. Denying that science had any special claim on the truth, Feyerabend wrote:

> Science is much closer to myth than a scientific philosophy is prepared to admit. It is one of the many forms of thought that have been developed by man, and not necessarily the best. It is conspicuous, noisy, and impudent, but it is inherently superior only for those who have already decided in favour of a certain ideology, or who have accepted it without ever having examined its advantages and its limits.[17]

If science does not have a special claim on the truth, or even on a unique ability to approach the truth, how are we to regard scientific statements and theories such as evolution? We may conclude, according to these philosophers, that they reflect far more the values and preferences of the ruling elite than the realities of the natural world—a world that we cannot, of course, know with any certainty.

For years there was little to fill the hollowed-out version of science that these philosophical attempts sought to create. To be sure, scholars in a growing field of "science studies" were happy to adopt the language of relativism, because it fit neatly into their own aims to study the scientific enterprise as a social and cultural phenomenon. They may have wished to shake up science a bit, to shatter the certainties of its most ardent interpreters, but they did not wish to displace it entirely. Intelligent design, however, had a more ambitious agenda.

Ironically, adopting the language of the academic left (Paul Feyerabend and many of the advocates of these views were indeed Marxists), intelligent design set out to argue that evolution occupied its privileged position in science education only by virtue of the fact that it was in keeping with the materialist ideologies of the ruling elites. Displace that ideology with another, equally valid system, and evolution would lose its status. This would be significant not only because it would make room for intelligent design as an explanation for the origin of species, but because it was Darwinian evolution, first and foremost, that had given the cover of scientific legitimacy to the philosophy of materialism. Strip that cover away, drive the wedge deep into the scientific establishment, and the whole rotten mass would collapse. The end result, according to the Wedge strategy, would be to reverse "the stifling dominance of the materialist worldview, and to replace it with a science consonant with Christian and theistic convictions."[18]

In short, by taking up the argument of the academic left that science could be deconstructed as a cultural phenomenon, the ID movement could not only win the battle against Darwin. It could win the greater war against science itself.

CASE STUDY

Kansas would be the first opportunity that ID had to place its stamp legitimately on the public school curriculum of an entire state, and its supporters went at the task with gusto. The election of 2004 had given antievolutionists a 6–4 majority on the Kansas Board of Education, and this new majority took up the issue of science standards early in 2005. What would follow was a remarkable demonstration of the ultimate goals of the ID movement.

Six years earlier a similar antievolution majority on the board had attacked evolution much more directly. Although they showed little interest in most of the science curriculum, in the summer of 1999 they removed evolution from the science learning standards to which students would be held. While board members were quick to point out that they had not actually outlawed the teaching of evolution in Kansas, much of the nation reacted as though they had. The goal of the board majority was actually to weaken the teaching of evolution, and by removing it from the list of scientific concepts that might appear on statewide exams, they surely had.

Unfortunately for them it was also a step that lent itself to caricature, and before long the state was the butt of jokes on television talk shows, and clever opinion columnists were making Wizard of Oz references so often that they became tiresome.

Much of the Kansas board had to stand for election in the summer of 2000, and evolution was the central issue in the campaign. It was not difficult for proscience candidates to ridicule the board for having made their state a laughingstock, and voters reacted accordingly. SCIENCE WINS IN KANSAS was the headline of a *New York Times* article following the election, and a new proscience majority reversed the board's decision on evolution. As satisfying as that might have been to proscience forces, the victory was temporary.

When a new group of antievolutionists was seated on the Kansas board early in 2005, they found that the process of writing new science standards for the state was well under way. A twenty-six-member committee had been working for much of 2004 under the leadership of University of Kansas Professor Steve Case, and they had already conducted a series of public hearings and prepared a detailed series of recommendations, which had the strong support of scientists and educators throughout the state, for submission to the board for approval. A minority of six members on the committee, however, including several well-known advocates of intelligent design, dissented from the committee's report and prepared a series of minority recommendations. These formed the basis of the new board's revisions of the science standards, revisions that were made final in December 2005.

The mistakes of 1999 had taught the ID movement a hard lesson in practical politics. They would no longer give their critics a chance to claim that Kansas had banned Darwin. In fact they proudly proclaimed that their new standards would require students to learn more about evolution than ever before, and in a sense that was correct. The problem was that much of what they were about to be taught simply wasn't true.

Part of the board's strategy was to weigh down evolution with philosophical baggage, such as these statements from the minority report:

> Biological evolution postulates an unpredictable and unguided natural process that has no discernible direction or goal. It assumes that life arose from an unguided natural process.[19]

By stating that the "theory of evolution postulates that change occurs through an unguided combination of chance circumstances and the operation of the physical and chemical laws alone,"[20] the board clearly sought to depict evolution in a way that would make it distasteful to as many Kansas students as possible. Another element of their strategy was to remind students as often as they could "that evolution is a theory and not a fact,"[21] a formula that, presumably, would undermine evolution's scientific status in the eyes of their students. But the board's most striking action was to attack something they called "methodological naturalism." The danger of this form of "naturalism," according to the Minority Report, was that it would promote a dangerous philosophy.

> The current definition of science is intended to reflect a concept called *methodological naturalism,* which irrefutably assumes that cause-and-effect laws (as of physics and chemistry) are adequate to account for all phenomena and that teleological or design conceptions of nature are invalid. Although called a "method of science," the effect of its use is to limit inquiry (and permissible explanations) and thus to promote the philosophy of Naturalism.... This can be reasonably expected to lead one to believe in the naturalistic philosophy that life and its diversity is the result of an unguided, purposeless natural process.[22]

When I first read this I couldn't quite believe my eyes. "Methodological naturalism" is not a philosophy; it is (just as its name suggests) a method. In a sense it is nothing more than the scientific method itself. The method, the technique, we use in science to try to answer questions is indeed naturalistic. When I see an unusual structure in the electron microscope, I try to identify and understand it using natural techniques, and I seek natural explanations for its form and function. When my car does not start on a cold winter morning, I also seek natural explanations for my problem, explanations that might include a dead battery, frozen gas line, lack of fuel, or bad wiring. But to the new standard-writers in Kansas, limiting science to natural explanations was far too narrow an approach. They sought to open science to nonnaturalistic explanations, and they didn't mean just evolutionary science—they meant *everything.*

The proposed changes to favor nonnaturalistic science were found throughout the new standards, but they were most striking in the definition of science itself. The original standards-writing committee had drafted its own simple, straightforward, commonsense definition:

> Science is the human activity of seeking natural explanations for what we observe in the world around us.[23]

It would be hard to improve on that definition, especially if the goal was to make it understandable to students at all levels, but the new board majority took issue with it. What they objected to, as they explained, was its "naturalistic" bias. Their new "less restrictive" definition read:

> Science is a systematic method of continuing investigation, that uses observation, hypothesis testing, measurement, experimentation, logical argument and theory building, to lead to more adequate explanations of natural phenomena.[24]

At first glance this might not seem so drastic a change, but there is a significant revision in the way the new board phrased its version. The list of terms like "experimentation" and "theory building" makes the new definition sound scientific, but while the original definition told students that scientists seek "natural explanations" for natural phenomena, the redrafted wording tells them that "more adequate explanations" will do just fine. Why "more adequate" instead of "natural"? The reason, as the board was careful to explain, was that confining science to natural explanations "philosophically limits both the formation and testing of competing hypotheses."[25] In their view it was absolutely essential to open science up to new, "nonnaturalistic" hypotheses. So complete was their fixation on "naturalism" that that word appeared no less than twenty-three times in the report upon which the new standards were based.[26]

The deeper danger of "naturalism," as the supporters of ID saw it, was that it carried an inherently antireligious message to the students of Kansas:

> Methodological naturalism effectively converts evolution into an irrefutable ideology that is not secular or neutral. Naturalism is the fundamental tenet of nontheistic religions and belief systems like Secular Humanism, atheism, agnosticism and scientism.[27]

In other words, in order to defend the science classroom against attacks from atheistic science, we simply have to change the definition of science itself.

When I first read the report produced to support the Kansas standards, I wondered precisely what its supporters envisioned that these new, nonnaturalistic hypotheses might be. Could they include the notion that travelers from space had visited the earth and given the knowledge to ancient civilizations to accomplish tasks like building the great pyramids of Egypt? No, because if extraterrestrial beings did exist, they would also be part of the natural world, so their actions would be naturalistic. Could the explanations invoke new principles or new laws of nature?

No, because such laws and principles would still be fully natural, even if we haven't discovered them. Try as I might, I couldn't think of a single nonnaturalistic explanation for anything that didn't involve a supernatural agent. And that, of course, was exactly what the board had in mind. They sought to make God, the great designer, a player, the solution to every unsolved problem in the history of life, maybe even in the history of the universe itself.

Gradually the nature of the board's not-so-subtle intentions became clear. Trying to avoid the difficulties that greeted the changes back in 1999, the board may have hoped to escape scrutiny this time around by leaving evolution in the curriculum. But their redefinition-of-science strategy was actually far more radical than the laughable measures of 1999, and would have not only prevented teachers from telling students that nonnaturalistic belief systems such as astrology, paganism, and wiccan healing are outside the realm of science, but would have undermined all of the natural sciences, not just biology, and certainly not just evolution.

Despite this tactic, once again Kansas voters got the message, and in the elections of 2006 they voted several of the pro-ID members off the board, returning it to a proscience majority. The bid to redefine science was reversed, and at least for the time being, students in Kansas will use the same definitions of science as everyone else. But the Kansas lesson serves as a warning of what may come to pass elsewhere.

BLOOM'S WARNING

The brilliant public relations efforts of the intelligent design movement have enabled it to attract support from broad segments of American society, especially including those who would identify themselves as political conservatives. In many cases the movement has successfully demonized evolution by depicting it as a liberal idea, a tool of the academic elite, a concept accepted only by those out of touch with the "traditional values" of American society.

As I have tried to explain in this chapter, there is a deep irony to the tactics employed by the intelligent design movement to marginalize evolution in American education. By adopting the arguments that Allan Bloom brought to bear against the academic left, they have sought to defeat evolution by introducing a new relativism into the practice of science. Once evolution is accepted as nothing more than the product of the ideology of "methodological naturalism," it will be easy to introduce another, theistic ideology in which intelligent design will qualify as equally valid science. If science is indeed just a myth enforced by the ruling elite, once that ruling elite becomes a truly Christian one, the science of intelligent design will validate its claims just as surely as evolution had validated the claims of the materialists. Such are the dreams of ID theorists, and such were their actions when they briefly gained control of education in Kansas in 2005.

Although Bloom is usually thought of as a conservative scholar, his conservatism grew out of a desire to preserve the Western ideal of scholarly inquiry as a search for genuine, objective truths in the humanities and social sciences. When a new generation of scholars stormed the bastions of academia under the banner of "openness," they challenged this traditional conception of learning, pushing it aside for an all-encompassing relativism that said, in essence, my truth is just as good as yours. There is no final answer to our questions, only relative ones constructed and shaped by the cultures in which we learn and grow.

Allan Bloom's concern was that the whole project of Western rationalism would fall to the onslaughts of cultural relativism, and this was his great fear for American higher education. It is not my purpose to judge the accuracy of his analysis or the timeliness of his concerns today, more than twenty years after the publication of *The Closing of the American Mind.* But I will acknowledge that I believe that Bloom was onto something, that he had grasped a valid truth about the nature of education and its effect on our society. To be open to all ideas and traditions is a fine thing, just as he admitted. But to allow that openness to keep one from a genuine, rigorous analysis of those ideas, to prevent competing ideas from being weighed and judged and discarded for cause, is to allow a false openness to paralyze the mind. The reason I find Bloom's ideas compelling in today's strife over the teaching of evolution is that it is exactly such a paralysis of judgment that is sought by the advocates of intelligent design. Science is a continuing exercise in rational judgment, an ongoing project to weigh competing ideas and to discard the ones of little or no value. Intelligent design creationism is just one such idea, but the essential strategy of its advocates is to prevent it from being discarded by turning Western science upside down for cultural, political, and religious purposes.

Am I being too extreme in my reaction to ID? Is it unfair to claim that ID's ultimate target is not to "correct" the mistakes of evolution but to destroy scientific rationalism itself? Consider these words by William Dembski, one of the acknowledged leaders of the ID movement:

> The implications of intelligent design are radical in the true sense of this much overused word. The question posed by intelligent design is not how we should do science and theology in light of the triumph of Enlightenment rationalism and scientific naturalism. The question is rather how we should do science and theology in light of the impending collapse of Enlightenment rationalism and scientific naturalism. These ideologies are on the way out... because they are bankrupt.[28]

There can be no mistaking the target in Dembski's crosshairs. To the ID movement the rationalism of the Age of Enlightenment, which gave rise to science as we know it, is the true enemy. The radical redefinition of science that was proposed in Kansas was only a hint of things to come. Science will first be redefined, and then the "bankrupt ideologies" of scientific rationalism can be overthrown once and for all. The modern age will be brought to an end, and the world will be the better for it, or so they say.

EIGHT

Devil in the Details

"What, in the name of Heaven, are scientists doing in a courtroom?"

That question was posed to me by one of my British colleagues at a scientific meeting more than a year ago. "Courts don't settle scientific issues. Science does that for itself, and it's best left on its own to do exactly that," he observed with just the slightest hint of condescension toward us unruly Yanks. He couldn't understand why I had found myself at a trial, explaining the science of evolution to a judge, or why we Americans tend to take such disputes to our courts with distressing regularity. I might have replied, "It's a First Amendment thing—you wouldn't understand," but I tried to give a more helpful answer. We hadn't gone to court to decide a scientific issue, I explained. Rather, we were in court to guarantee that science could continue to do exactly what my friend thought best—to allow the scientific process to work without interference. The danger, I told him, wasn't that science would get things wrong. The danger was that politicians would decide that they could tell science what was true and what was false, and would dictate that a bogus version of science be presented to American students. That, I assured him, might even get an Oxford grad unsettled.

Or maybe not was the look he shot back at me. *Because it wouldn't happen in* my *country.*

Maybe not.[1] But it certainly is happening here. Nearly every American would agree that evolution is a "controversial theory" because they have seen firsthand the strife that it produces. Whenever I show up on a radio talk show and the topic of evolution is announced, the switchboard lights up immediately. Everyone has something to say about the subject. Some are critical, some are supportive, but all opinions are strong ones, deeply held, and expressed with considerable passion.

In one sense this is easy to understand. Evolution affects each of us in ways far more personal than atomic theory or the germ theory of disease. Evolution speaks directly to our conception of who we are, where we come from, and how we should regard ourselves with respect to the rest of the living world. Like it or not, evolution hits us right where we live. That's one of the reasons why it provokes such strong reactions. The other, as we've seen, is because many regard evolution as the cutting edge of a dangerous and destructive movement—a drive to secularize society and to undermine the traditional values that they believe have built our country. For many Americans, if evolution threatens the moral foundation of society, the issue of whether it is scientifically correct is secondary. It's a dangerous idea, and that's all that matters.

Given the amount of controversy evolution has engendered, what should we do? If science education is genuinely important, and most Americans would agree that it is, why not just soft-pedal the controversial stuff, and concentrate on the scientific topics that don't make sparks fly? Kathy Cox, Georgia's superintendent of schools, actually tried something like that approach in 2004. Noting how every occurrence of the word "evolution" in the state standards provoked hostile public comment, she had the standards rewritten in a way that avoided the E word entirely. When its absence was noticed she defended her actions by explaining that students would still be asked to study the same material—they just wouldn't call it "evolution." Perhaps she hoped that nobody would notice what he or she was actually studying, and the business of science education could go on in peace. Needless to say, that didn't happen. The E word went back into the standards, and it remains a flashpoint in Georgia and just about every other American state.

Facing such hostility and resistance, why not just give up? There's plenty of biology to be learned outside of the lessons on evolution. Would it really be a tragedy if students missed out on allopatric speciation and the Hardy-Weinberg equilibrium (big topics in evolution) and concentrated instead on food chains, action potentials, and DNA replication? Of course not. There are always gaps in education, and bypassing a few topics, even important ones, is hardly going to bring down the whole enterprise. In science we're fond of emphasizing that understanding process is a lot more important than memorizing facts, so if a few facts are missing, what's the harm?

More than a few of my colleagues in science take a similar attitude when they hear about resistance to the teaching of evolution in one state or another. As one of them remarked to me a few years ago, "Who cares what they teach kids in Alabama or Mississippi? We aren't getting many scientists from those places anyway, so let 'em teach whatever they want."

Having just returned from leading a teacher workshop in Jackson, Mississippi, I was tempted to get on my high horse and lecture the guy. I figured it might be good to remind him that E. O. Wilson, the great Harvard evolutionary biologist, grew up in Alabama, and for all he knew, the next E. O. Wilson or Stephen Jay Gould was growing up there right now. But I figured that his geographic contempt for certain regions of the country was so profound that my protest wouldn't have made any difference.

Nonetheless it's only fair to ask if we should insist on making a big deal about evolution. It is, as its opponents keep reminding us, "only a theory," so why not take a deep breath, step back from the brink, and stop demanding that it be taught? There's already plenty to learn in school. It wouldn't substantially handicap history students if we skipped the Gadsden Purchase, nor would it prevent our kids from understanding English if they missed the definition of a gerund, would it? So why not skip Darwin's theory, or at least its more troublesome elements, and see if that will bring us peace?

The problem with this argument is that evolution isn't "just a theory"; it's the glue that binds the biological sciences together, a common principle that links fields as diverse as development, genetics, and paleontology. In an age in which the flow of life science research has moved toward unification around evolution, taking Darwin out of the picture would send biology backward into an age of fragmentation. But the biggest problem posed by avoiding evolution is much greater than that. In a very important sense evolution is the canary in the mineshaft, an indicator whose presence signals the health or sickness of the entire scientific enterprise. That, ultimately, is the point of this book. The question of evolution is really a question of what will happen to the American soul.

REACHING FOR THE GRAIL

Conspiracy theories are a reliable staple of books and movies, and it's genuinely tempting to dismiss the ID movement as the result of just such a conspiracy. All of the basic elements are there, including a semisecret founding document ("The Wedge"), behind-the-scenes machinations by the wealthy and powerful, and confidential meetings to coordinate the efforts of true believers. The ID movement was, in fact, the deliberate creation of a small group of individuals, whether you wish to regard them as conspirators or visionaries, and much of its early work was carried out well hidden from public view.[2]

But such an analysis, whether accurate or not, misses the point. Intelligent design, regardless of its origins, has struck an extraordinarily responsive chord with the American public. If the movement's claims are correct, if its science is sound, and if its message is appealing, why would it matter if it was cooked up by a few smart guys in a back room? The real question isn't where ID came from, but where it wants to go.

So, what does the ID movement really want? What are its genuine goals? Many who find the movement attractive insist that all they are really asking for is an honest appraisal of the evidence for and against the theory of evolution. This is not possible, they insist, so long as "Darwinists" maintain their stifling control over education and free inquiry. If this view were correct, the real goals of such a movement would be modest indeed. They would amount to nothing more than ensuring that the scientific process had a chance to work and to put evolution to the same sort of examination to which all scientific ideas are subjected. And what could be the harm in that?

No harm at all, I'd be quick to say. In fact that's exactly the sort of scrutiny that Darwin's theory has always faced. Scientists are a tough crowd, and they continue to put evolution to the test. But the goals and tactics of the ID movement go far beyond simply getting a fair hearing against evolution. What they seek, as explained in the Wedge document, is nothing less than the overthrow of materialism and its cultural legacies in favor, as Phillip Johnson put it, of a "theistic science," a new kind of science that would use the Divine not as ultimate cause, but as scientific explanation. Scott Minnich, a University of Idaho microbiologist and Discovery Institute fellow, made exactly this point at the Dover ID trial, noting that for ID to be considered science, "the rules of science have to be broadened so that supernatural causes can be considered."[3] Minnich didn't restrict this "broadening" of science to the question of the origin of species or of life itself. Steven William Fuller, another ID witness at the trial, made it clear that the principal aim of the ID project is to "change the ground rules of science to include the supernatural"—a project that was briefly achieved in Kansas when ID supporters struck any suggestion that science must be limited to natural causes from that state's education standards.

The supporters of ID like to claim that their intentions aren't all that radical and that they simply want to return to a traditional definition of science, one that would be understood by the likes of Newton, Galileo, and even Einstein. These scientists regarded their work as revealing the greater glory of God by exploring his handiwork, so clearly their worldviews included (nonnatural) causation, just as ID tries to do today. But that argument fails to appreciate the clear distinctions that each of these scientific giants drew between their scientific projects and their religious beliefs. Einstein may have speculated frequently on how God might have structured his universe, but he never wrote the explanation of God into an equation, or attributed an experimental result to the actions of the supernatural. Newton may have felt that the orderly mathematics of nature implied a divine lawgiver, but he relentlessly sought natural causes for every natural phenomenon he investigated, from the movement of the planets to the nature of light. Indeed, if there is one characteristic that has distinguished Western science from every form of inquiry in human history, it is its uncompromising insistence that nature itself must be the source of answers for questions about the natural world. That's the foundation not only of the Western post-Enlightenment tradition, but of science itself.

What would happen to science if its ground rules were changed? What would a science of the future look like if we considered "nonnaturalistic" causes to be legitimate scientific explanations? At a stroke they would be accepted in every branch of science. That earthquake devastating part of the third world might have been caused by the shifting of tectonic plates, but it could also be a punishment for the sinfulness of those now suffering in the rubble. Why bother to conduct an exhaustive molecular search through simian virus genomes to find the source of HIV when clear-thinking ID scholars have concluded that it was sent as a divine warning against deviant lifestyles? In fact even the rainbow might just be a phenomenon presented to us by a "whimsical" designer, according to ID theorist William Dembski.[4] Why worry about the physics of light when the mystery of the rainbow can be solved by easy reference to the personality of the creator?

Once the supernatural becomes a valid element in scientific inquiry, science will cease to be an empirical search for the truth of the natural world. Like faith itself, "theistic science" will be a subjective window on the world that reflects the innermost convictions of its adherents and not the outer reality of nature, the stringent standard by which speculation and hypothesis are forged into scientific theory. A theistic science may be friendly to the tenets of faith, or at least to the faith of some, but it will no longer be the science we have known. It will cease to explore, because it already knows the answers. And humankind will be the poorer for it.

DESTROYING THE VILLAGE

It is possible, of course, that ID will never achieve this level of success, and that the project of a theistic science is fundamentally so absurd that no sensible society will ever embrace it. That may turn out to be true, but it would be cold comfort indeed if the tactics of the ID movement, even if they fall short of their ultimate goal, leave the scientific enterprise in tatters. And they just might.

Consider for a moment the breadth of the attack that has already been mounted against evolution by the ID movement. Although the warriors of design consider themselves to be scientists, they argue that science as we know it simply is not working. The self-checking mechanism of scientific competition no longer operates, according to their account, because a stifling orthodoxy has seized control of the scientific community. Dissent is punished, orthodoxy is enforced, and progress is systematically prevented. With such careful and well-placed rhetorical attacks, one can depict the most open and self-critical of all human enterprises as narrow and close-minded. This has indeed been the key tactic of the ID movement, aptly described by Barbara Forrest and Paul Gross:

> Here is how you implement it: exploiting that modern, nearly universal liberal suspicion of zealotry, you accuse the branch of legitimate inquiry, whose results you hate, in this case the evolutionary natural sciences, of—what else?—zealotry! Fanaticism! Crying "viewpoint discrimination," you loudly demand adherence to the principle of freedom of speech, especially in teaching, insisting that

> such freedom is being denied your legitimate alternative view. You identify your (in this case, religious) view of the world as a victim of censorship by a conspiracy among the world's scientists, whom you label "dogmatic Darwinists" or the like.[5]

Once the movement has made its case that science is no longer open and self-correcting, the door can be opened to "corrective" measures from the outside. Those measures, of course, include political efforts to "fix" science by ensuring that "alternate theories" are given equal time in the classroom and equal attention in the curriculum. The rallying cries of "fairness" and "balance" sound a lot more palatable than "political interference," but they amount to the same thing. Every time a legislature or school board or education commissioner calls for balanced treatment or critical analysis of evolution, the tactic has scored another success. Despite a few recent reversals in court and at the ballot box, there is every sign that this maneuver will continue to work, and that scientists and educators will face continuing pressure from the political sphere to "soften" the teaching of evolution and to introduce alternatives that have almost no support in the scientific community.

If those in the ID movement continue to find success with this approach, they may indeed gain a hearing for their viewpoints among America's schoolchildren. But I wonder if even they have considered the magnitude of the indirect damage they will have done to science. It is one thing to get their ideas into the heads of schoolkids, but in the process they are changing the very conception of what science is and how scientific judgments are made. They may actually believe that their ideas are correct and therefore they should be advanced by any means necessary, but consider what such successes would convey to Americans about the nature of science.

First, they send the message that the scientific process is not to be trusted, and not only because science is a human activity and as such frequently makes mistakes, even big ones. The hidden message is that science is no different from any other political activity in which the opinions, prejudices, and viewpoints of the dominant groups are reinforced and maintained at every level. In other words, science isn't science; it's a way of bending the stories we tell about nature in order to support a larger political goal. In the case of Darwinists it's a story of a pointless, meaningless existence supporting the politics of secular materialism. In the case of ID the story is one of meaning, purpose, and value that endows every human being with the special dignity accorded the intentional creation of a supreme being. Which one do you choose? Whichever you prefer. The choice is one of political philosophy, not of the empirical reality of a natural world in which science can be used to provide a genuine test of alternatives, and to rule out the ones that don't fit the facts.

Second, ID proponents set a new model for how science is done. Once you've developed an innovative idea, a concept (like intelligent design) that you feel has genuine value, what's your next step? Apparently it's not to do serious research that puts that idea to a genuine test. Rather, it's to engage in public relations and political activity designed to gather a strong constituency in support of your idea. Once you've accumulated political strength, then it's time to do an end run around the scientific community by appealing directly to agencies of government to inject your ideas into classroom and curriculum by force of law. To be sure, the enthusiasts of design are certain their ideas are valuable and correct, and therefore they see nothing wrong with sidestepping the pesky process of peer review and skipping the time it takes to win a consensus within a skeptical scientific community. But if they are even modestly successful, the example they have set will become a permanent one for how "science" will be practiced in the future.

Finally it's only fair to ask what will be left of the scientific community if the tactics of ID continue to work so well. The profound meritocracy of science will vanish in an instant if the ID movement is successful—not because the doctrines of ID are so threatening, but because the way in which it seeks to wedge itself into mainstream science is completely alien to the scientific culture. I don't doubt for a minute that the advocates of ID think that would be a fine development, because they don't much like the scientific culture to begin with. But the distinguishing feature of the scientific enterprise, from its earliest beginnings, has been the power of an idea to overcome popular resistance if it provides a better description of the natural world. The notion that anyone, whether part of a powerful elite or not, can revolutionize scientific understanding with a novel idea has been validated time and time again in scientific history. This certainly doesn't mean that such ideas are immediately accepted or that criticism is always warmly welcomed, but it does mean that the ultimate test is found in nature itself—not in the prevailing politics of the day. Change those rules, replace the scientific culture with a political or religious one, and science will slip away. Its most passionate practitioners, its most creative minds will recognize that the core values of the enterprise have vanished, and they will find other fields in which to exercise their talents.

In the language of America's ill-fated war in Vietnam, the warriors of ID have concluded that they must "destroy the village in order to save it." The scientific "village" may not be infested with Vietcong, but it's packed to the rafters with secular humanists, so nothing less than a complete assault will do. The partisans of ID are lobbing intellectual napalm into the scientific community, and so great is their enthusiasm for their tactical objective that they remain oblivious to the fact that nothing will remain but ashes and dust if their attack is successful. They are happy to fire missiles of doubt and relativism to storm its walls, but make no mistake about the ultimate effects of these weapons: The village as we know it will burn to the ground, and the qualities that made science such a valuable prize will be lost—perhaps for a very long time.

THE POLITICS OF DESIGN

For me, one of the great attractions of science has always been its profoundly nonpolitical nature. Scientific arguments can get heated, and scientists certainly play the game of personal politics in their institutions, journals, and societies, but the real-world politics of government and political philosophy play a very little role in science. Partisans of both the left and right resist this view, I suspect, because it limits their own ability to enlist science in support of their respective agendas. But it's true nonetheless. At the very height of the cold war I struck up a brief collaboration with a group of plant biologists in East Germany who were able to provide me with some unique specimens that promised to aid my research. At first we dealt with one another only by letter (this was in the pre-Internet age of the 1970s), but eventually I met several of my collaborators at a plant science conference in Greece. As we discussed our work, joking about who was having more fun opening our mail, the Stasi or the CIA, we compared the conflicting views that our different nations held on history, art, and even music. The one thing, we agreed, that wasn't tied up in the ideological struggle between East and West was science. Both sides might try to use science for their own ends, but in the end photosystems I and II were the same on both sides of the Berlin Wall. As scientists, we found it easy to work together. However one tried to twist it, science resisted all attempts to make it political.

To most Americans this is not the case with evolution. A few years ago I asked about a hundred of my freshman biology students where they would place evolution on the political spectrum. Was it a right-wing idea, was it middle of the road, or was it left-wing? Nobody placed it on the right. A fair number raised their hands for middle of the road, but most of my nonrepresentative slice of young America associated evolution with the political left.

This is certainly true of the American public as a whole. Americans who call themselves conservative[6] reject evolution at much higher percentages than self-identified liberals. The antievolution movement consistently identifies evolution as politically leftist (or even communist) and sees Darwin's theory as the cutting edge of dangerous social trends that act against the traditional, conservative foundations of American law and society. The intelligent design movement, by contrast, is frequently embraced by conservative thinkers, and not just because of the convenient alliance it helps to cement with the religious right. When they see the support that evolutionary science enjoys in America's great research universities and colleges, and then consider the leftist bent of much of the political rhetoric coming from such places, they know exactly what choice to make. Reflexively they stand behind the only alternative they can—intelligent design.

The curious aspect of this politicization is that if evolution does have any message for politics and economics, it is almost entirely in the other direction. My earliest glimpse of this was back in high school, when one of our teachers led a class trip to visit the stock exchange in nearby New York City.

The first time I saw the exchange in operation from our perch in the visitors' gallery, I simply couldn't believe the chaos on the trading floor. "Isn't there anybody in charge?" I muttered out loud. How could a system where everybody was only out for number one achieve what my teacher assured me was the world's most efficient allocation of capital? Wouldn't the best possible system come from careful, intelligent planning? Wouldn't the best result come from "design" instead of the confusion of the trading floor? Not for the first time, I quietly concluded my teacher was a dope.

As Mark Twain might have said, it's amazing how much smarter he's gotten in the last forty years. The decades since my high school years have seen the world's planned economies collapse in dismal failure, while those that relied on the "chaos" of the marketplace have surged ahead. There's nothing new in the ultimate success of unplanned economic activity, of course. Centuries ago Adam Smith described the surprising efficiencies that spring from the uncontrolled self-interest of the individual: "He generally indeed neither intends to promote the public interest, nor knows how much he is promoting it. . . . He intends only his own gain, and he is in this, as in many other cases, led by an invisible hand to promote an end which was no part of his intention."[7]

Capitalism, as conservatives never tire of pointing out, produces economic efficiency not by design from above, but from innovation, investment, and self-interest from below. The ability of modern capitalism to invent, adapt, and prosper stands as dramatic testimony against those who would argue that complexity and efficiency cannot arise spontaneously, but must be planned into a system by a supervising authority. Charles Darwin would have loved it.

What impressed Darwin, as well as many others, about living things was how well-suited they are to their environments. Other naturalists could do no better than to attribute this to careful, centralized planning, but Darwin knew better. He supplemented his observations on natural systems with studies of the economic theories of Thomas Malthus and Adam Smith, whose work preceded his by a generation. From economics he gained one of the key insights of his theory: namely, that allowing individuals to struggle for personal gain helps to weed out inefficiencies and produces a balanced system that ultimately benefits society as a whole.

In a certain sense Darwin's theory of evolution by natural selection is unadulterated Adam Smith translated into the language of biology. The unthinking acts of individual organisms, seeking no more than survival and reproductive success, produce biological novelty just as surely as venture capitalists foster innovation.

At the beginning of the twenty-first century evolution has won the scientific argument just as surely as capitalism has won the economic one. Ironically, the critics of evolution, who otherwise fashion themselves as conservatives, are the ones who argue for central planning and design in the organization of living systems. By doing so they fail to acknowledge Smith's great lesson that the most intelligent design of all, whether in economics or biology, comes from having the wisdom to let nature work its course.

The truth is that if Charles Darwin were to appear today in midtown Manhattan, I know exactly where I'd take him first. No, it wouldn't be up to the Museum of Natural History, whose rich collections of fossils have so eloquently documented the historical details of evolutionary change. It wouldn't even be to the great university laboratories, where studies of molecular genetics have provided the mechanisms to support his theories. It would be to a place where people would *really* understand him, a place where his theories are put into practice every day, a place where a true evolutionist can have a rip-roaring good time. I'd take him to Wall Street, and then I'd tell him about that brilliant teacher I had in high school.

For some conservatives these lessons are all too obvious. Not only should economies be left alone to produce the efficiencies that lead to greater prosperity and individual freedom, but science should be left alone to do its thing as well. A science that is "corrected" or "balanced" by the political process is no longer science. The scientific community, more than any other institution in our society, is a free marketplace of ideas where only the fittest prevail, and the unwillingness of ID to subject itself to competition in that marketplace speaks volumes.

George Will, one of the genuine leaders of American conservatism, recognized this fact in a column written for the *Washington Post* after voters in Dover, Pennsylvania, turned their pro-ID school board out of office in 2005. The election results, according to Will,

> expressed the community's wholesome exasperation with the board's campaign to insinuate religion, in the guise of "intelligent design" theory, into high school biology classes, beginning with a required proclamation that evolution "is not a fact."
>
> But it is. And President Bush's straddle on that subject—"both sides" should be taught—although intended to be anodyne, probably was inflammatory, emboldening social conservatives. Dover's insurrection occurred as Kansas's Board of Education, which is controlled by the kind of conservatives who make conservatism repulsive to temperate people, voted 6 to 4 to redefine science. The board, opening the way for teaching the supernatural, deleted from the definition of science these words: "a search for natural explanations of observable phenomena."

> "It does me no injury," said Thomas Jefferson, "for my neighbor to say there are twenty gods, or no God. It neither picks my pocket nor breaks my leg." But it is injurious, and unneighborly, when zealots try to compel public education to infuse theism into scientific education. The conservative coalition, which is coming unglued for many reasons, will rapidly disintegrate if limited-government conservatives become convinced that social conservatives are unwilling to concentrate their character-building and soul-saving energies on the private institutions that mediate between individuals and government, and instead try to conscript government into sectarian crusades.[8]

To another conservative thinker, Charles Krauthammer, the religious impulse to embrace ID is just as foolish. Also writing in the *Washington Post,* Krauthammer called ID a scientific "fraud," but he also made the more expansive point that Darwin's great idea was anything but contrary to faith.

> How ridiculous to make evolution the enemy of God. What could be more elegant, more simple, more brilliant, more economical, more creative, indeed more divine than a planet with millions of life forms, distinct and yet interactive, all ultimately derived from accumulated variations in a single double-stranded molecule, pliable and fecund enough to give us mollusks and mice, Newton and Einstein? Even if it did give us the Kansas State Board of Education, too.[9]

Conservatives view themselves as the guardians of tradition, especially of a Western intellectual tradition that produced the liberal democracies in which science and free enterprise have flourished side by side. The familiar enemies of that tradition, promoting a collectivist approach to politics and economics, denounce the results of the free market and argue that central planning and state control are the solutions to society's problems. Conservatives rightly view these critiques as subversive to their intellectual heritage, and have set themselves against movements that call for economic and political socialism.

Curiously, in the case of ID, many still fail to see the devil in the details. The ID movement poses exactly the same collectivist threat to Western rationalism, differing only in the detail that it threatens academic science and science education instead of the economic and social order. Decades of bashing academia have clearly made it easy for conservatives to oppose "intellectual elites," including the scientific one, and the ID movement has cleverly capitalized on this tendency. A closer look, however, shows that the antievolution movement is truly radical in its character, aimed at using political power to distort the scientific marketplace to achieve a predetermined end—the endorsement of a particular religious view of nature. A true conservative wants to let that marketplace continue to function as it has for centuries. A true conservative is on the side of science.

ID'S GETTYSBURG

Thinking of the role that Kansas played in the years before the Civil War, it seems only fair to wonder if history has repeated itself. Electoral politics in Kansas, as we've seen, have recently made it a burned-over district in the struggle over evolution. A century and a half ago the issues that divided Kansas went national, and so they have again. And, very much as in the Civil War, the struggle reached a climax, if not a resolution, in a small town in Pennsylvania.

History may one day record that the pivotal battle in ID's war against evolution began, much as the battle of Gettysburg did, with an almost accidental skirmish. In 2004 the science department at Dover Area High School in Dover, Pennsylvania, was allowed to choose new textbooks for its general biology classes, and its members selected a book, *Biology,* published by Prentice Hall and written by Joseph Levine and me. The first hint of trouble came when members of Dover's board of education balked at approving their choice. One of the board's members complained that the book was "laced with Darwinism from beginning to end"[10] and set about helping to present an alternative to teachers. The board also arranged for the purchase of two classroom sets of the ID textbook *Of Pandas and People,*[11] which were placed in the high school library. What followed led to a First Amendment trial on the issues of evolution and intelligent design, drawing worldwide press attention to the small town of Dover and to the continuing battle over science education in the United States.

On December 14, 2004, a group of eleven parents of students in the Dover School District filed a lawsuit in federal court alleging that the Dover Board of Education had violated their constitutional rights. *Kitzmiller v. Dover,* as the lawsuit is known, charged that by using government power to bring the idea of intelligent design into public school classrooms, the board had, in effect, established a religion in violation of the First Amendment to the Constitution. The case moved toward trial with remarkable speed, and on September 26, 2005, Judge John E. Jones III called both sides to order in federal court in Harrisburg, Pennsylvania. I was called as lead witness for the plaintiffs and spent much of the first two days of the trial in the witness stand, making the case for evolution and being cross-examined by attorneys defending the school board.

The details of the trial have been widely reported, and a complete recounting of the proceedings has been published in a number of other books and articles.[12] Suffice it to say that, as the trial played out, events in the courtroom provided for exactly the sort of grand confrontation for which many partisans in the struggle had been hoping. Only four months earlier William Dembski had fantasized about the opportunities that such a trial might present for ID:

> I therefore await the day when the hearings are not voluntary but involve subpoenas that compel evolutionists

> to be deposed and interrogated at length on their views. There are ways for this to happen, and the wheels are in motion.... What I propose, then, is a strategy for interrogating the Darwinists to, as it were, squeeze the truth out of them.[13]

Lest there be any doubt as to Dr. Dembski's seriousness in this matter, he actually included with his blog entry containing this text photographs of a Charles Darwin doll with its head placed firmly in a vise. Clearly the Dover case was even then shaping up to be the decisive battle in which the ID movement would prove itself.

But a funny thing happened on the way to the trial. While Dr. Dembski and seven other proponents of intelligent design signed on as expert witnesses to defend the Dover board, only three of those eight actually appeared in court—and Dembski was not among them. Citing conflicts of interest with their publishers, and demanding personal counsel during their depositions (in addition to the lawyers representing the Dover board), one by one most of the experts favoring ID declined to appear in court. Perhaps they feared what was about to transpire.

What did take place in the courtroom was, by any standard, a scientific rout. As noted earlier in this book, key arguments of the ID movement were examined, and one by one they collapsed, often in spectacular fashion. Claims of peer review for an ID book were shown to be not quite what they seemed, and the "research" produced by ID proponents clearly did not meet any reasonable definition of science. Attorneys for the board, as part of their defense, had promised they would show that ID met the legitimate standards for a scientific theory. To say that they failed in this regard would be almost too kind, and this is but one of many lessons from the Dover trial. The others included what the judge described as the "breathtaking inanity" of the Dover Board of Education, the persistent attempts of the ID movement to obscure its religious roots, and the willingness of self-professed people of faith to come into his courtroom and lie under oath.

Since the trial's conclusion the advocates of ID have struggled mightily to downplay its significance and even to besmirch the reputation of the trial judge, a conservative Republican appointee. They have tried to argue that the judge's opinion was plagiarized from material provided by the ACLU, that the Dover Area School Board rejected their pretrial advice, and that witnesses such as myself "mischaracterized" intelligent design and "misrepresented" the ideas of ID proponents such as Michael Behe.

The Dover board may indeed have been poor students of the strategies promoted by the Discovery Institute, but in retrospect it is clear that the board's primary failing lay in being too honest about its own goals and motivations—one of its members had actually been captured on videotape advocating the teaching of creationism to "balance" evolution, and others had made similar comments at open meetings. Under such circumstances it was easy to demonstrate that the board members' primary intention had been to advance their own religious point of view.

If I or other witnesses actually had misrepresented ID or the ideas of its proponents, those misrepresentations would have been easy to correct. Since the defense (the Dover board) presented its case after the plaintiffs, Michael Behe's testimony followed mine. Behe's principal claim, as we have seen, is the argument that biochemical systems are "irreducibly complex" and therefore must have been put together by an unnamed designer. Much of the evidence I had presented in my testimony to counter this claim was summarized in chapter 3 of this book, and it offered a clear challenge to which Behe had every opportunity to respond. Despite this, in three days on the witness stand, Dr. Behe failed to convince the judge, or for that matter anyone else in the courtroom, that ID had been unfairly maligned or inaccurately presented to the court. Quite the contrary. He actually convinced nearly everyone present that the "irreducible complexity" argument was a hopeless failure. The judge's opinion tersely summarizes the sorry state of the ID argument after Behe's testimony:

> We therefore find that Professor Behe's claim for irreducible complexity has been refuted in peer-reviewed research papers and has been rejected by the scientific community at large. Additionally, even if irreducible complexity had not been rejected, it still does not support ID as it is merely a test for evolution, not design.[14]

No doubt there must have been a few people back in 1863 who were willing to cast the Confederate defeat at Gettysburg as nothing more than a tactical redeployment. Lee's army did survive to fight another day, and the outcome of the war remained in doubt. The same is true of the ID movement after Dover. The precipitate withdrawal of most of the Discovery Institute's expert witnesses helped to limit the level of personal damage to the "stars" of design, and their forces may yet win the victory they seek—to change the nature of science. The outcome of the evolution war remains in perilous doubt.

Nonetheless, the parallels between Dover and Gettysburg remain. To date, Dover represents the high-water mark of the ID movement just as surely as the Confederate advance into Pennsylvania set that mark for Lee's army. The "capture" of a school district, and its embrace of ID, were the first time that a local board had actually ordered its science teachers to bring this upstart into the classroom. ID's enthusiasts surely hoped that this advance would break the Darwinist stonewall much as Pickett's charge was designed to send Meade's army into retreat. Instead, the Union forces held, and Pickett's soldiers were bloodied and crushed. Lee had bet that his forces could defeat the enemy in a single head-to-head battle, but he lost that bet. At first the strategic forces of ID had thought much the same thing: Win the battle of public opinion and ID would demonstrate its scientific superiority in any forum, even the courtroom. For them the defeat was total, the outcome bitter, and the future they sought attainable only by denying they had wanted the battle in the first place. That is exactly the position they have taken since the Dover trial, and will continue to take—until the time is right, at some point in the future, for another frontal assault.

THE WORST OF TIMES

Exploiting the tension between extremes is a dramatic device familiar to every actor, every playwright, and every author of fiction. The reason it works, of course, is because it reflects real life—the conflicts we feel within ourselves are often between extremes, and the resolution of those conflicts can be sudden and dramatic. Two indifferent strangers fall suddenly in love, an alcoholic turns away from his addiction, a hesitant son unexpectedly finds the courage to confront his murderous stepfather. The choice between extremes is the stuff of genuine drama, and as readers we connect with it instinctively. No writer exploited this principle more often or more effectively than Charles Dickens, and his masterpiece *A Tale of Two Cities* is literally built around the framework of conflicting tensions.

Early in the book the reader is presented with an unremarkably ordinary scene on a busy French street. People bustle past one another on crowded sidewalks until something happens that shocks them from their lonely solitude: A cask of wine bursts and spills its contents on the street. In a flash the passersby wake from their collective trance and become greedy scavengers. They dash for the pools of spilled wine, drinking it out of their bare hands and scooping it into anything handy, even their own cloaks and bandanas. And then, almost as suddenly, the frenzy ends when all of the red liquid has been consumed. The message is unmistakable: The people of France are ready to snap, and the capacity for sudden violence is in each of them, however they may try to repress it. When the wine is gone, a person whom Dickens describes as a "joker" scrawls a word upon a wall using the reddish tint of the wine, and the word he writes is BLOOD. The desperate tension is not only palpable, but strong enough to drive a revolution—a revolution tinged in blood.

Charles Dickens began his narrative with one of the most memorable lines in literature: "It was the best of times, it was the worst of times." And so it was. He set his drama of the French Revolution against the contrasts between London and Paris. "It was the spring of hope, it was the winter of despair." The fate of the revolution to come would teeter on the sharp edges of idealism, oscillating between humanity and depravity, between tyranny and liberty. "It was the season of Light, it was the season of Darkness." Events were driven by the noblest aspirations of humanity and yet were always in danger of awakening our most savage instincts. Stability and certainty were the first casualties of the movement toward revolution, and the core of Dickens's remarkable novel concerned a series of conflicting intrigues, personal and political, that could have turned either way. In every case it could easily have been the best of times or the worst of times. The dance of fate between two extremes would decide the futures of governments and nations, of men and women.

To many of my scientific colleagues, the image of science in the United States suggests that we are already in the worst of times. The public opinion of science as a profession is still fairly high, but it has declined steadily for nearly two decades. The media stereotype of scientists as a bunch of hopelessly out-of-touch geeks is very much alive and well, and there seem to be remarkably few spokesmen willing to take the case of science to our popular media. As one of my colleagues remarked, "Where is Mr. Wizard, now that we need him?"[15]

Evolution is routinely rejected by more than half of the American people, who defiantly tell pollsters that they regard the very core of the biological sciences as a fundamentally erroneous concept. A recent poll of attitudes toward evolution in Japan and thirty-two European countries demonstrated how exceptional the American rejection of evolution truly is (figure 8.1).[16] In this survey Americans stood next to last in their acceptance of evolution, above only the citizens of Turkey. If you're an advocate of creationism or intelligent design, of course, you might well regard these as the *best* of times. After all, as one of my scientific friends quipped, when ID wins, we'll automatically be in second place.

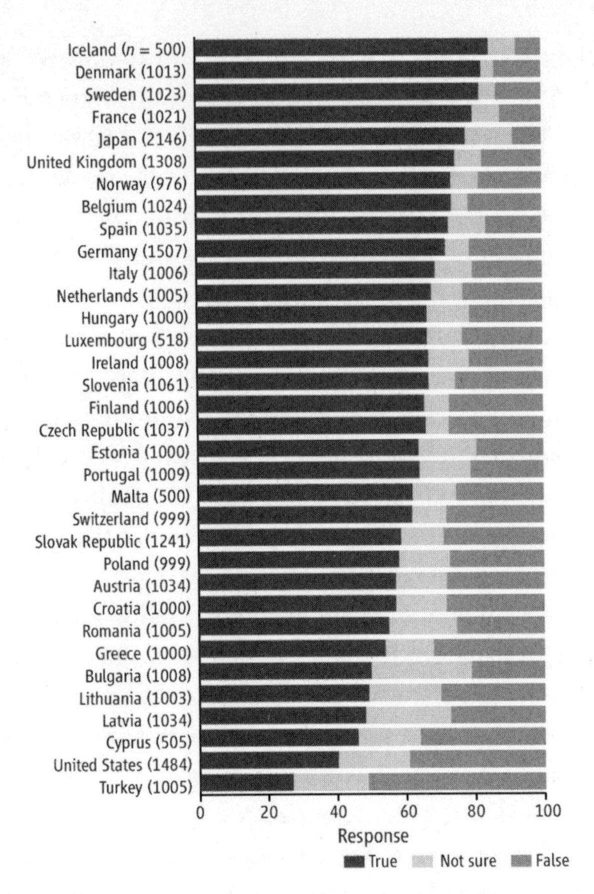

Figure 8.1: Public acceptance of evolution. A survey in 2005 showed that the United States is near the bottom among industrialized countries in the percentage of public acceptance of evolution. Respondents in thirty-two European countries and Japan were given this statement: "Human beings, as we know them, developed from earlier species of animals." They were then asked if the statement was true or false, or if they were not sure. Numbers in parentheses are the numbers of individuals questioned in each country. *(J. D. Miller et al., "Public Acceptance of Evolution in 34 Countries,"* Science *313 [2006]: 765–66. Reprinted with the permission of AAAS.)*

Setting that little glimmer of optimism aside for a moment, how should we regard the current state of American science? If we were to look only at what Americans think about evolution, the news could hardly be worse. Most Americans seem to think that evolution is nonsense, and they find it far more reasonable to believe that human beings were created, in pretty much their present form, only a few thousand years ago. In a June 2005 Harris poll,[17] only 38 percent of Americans thought that human beings had developed from an earlier species, while fully 54 percent thought that we had not. In other words, they believed that we were directly created. If evolution is indeed the cornerstone of modern biology, how can America consider itself a modern scientific nation when a majority of its citizens reject that cornerstone as unsound?

One might well spend considerable time worrying about the consequences of these findings. By rejecting evolution we reject the notion of common ancestry, denying our kinship with other forms of life, and perhaps making it more difficult to argue the case for the preservation of biological diversity on our planet. By denying that mutations can be beneficial or creative, we close our eyes to the evolving challenges of infectious disease, failing to see that evolution is the key to dealing with the changing armies of microbial pathogens. By contending that the long history of life on this planet is mere illusion, we ignore the lessons of paleontology—that the history of life is a history of extinction and change, and that we have no reason to believe that our presence here is any more permanent than that of the species that preceded us. We ignore such practical lessons at our peril. But the real dangers of such willful rejection of the lessons of science come at a much deeper level.

I began this book by discussing the state of science in America and wondering why the scientific enterprise had found such fertile ground in our democracy. It certainly is not because the United States has given its children the best scientific education in the world—you have to look long and hard to find an educational category in which we're even among the top ten. It certainly is not simply because we devote more money to science than other countries, and it's surely not because our people have some special genetic makeup to incline them scientifically. We Americans carry the genes of every nation in our collective DNA. It is, I have argued, because there is something in the American ideal of individualism that resonates with science, that rewards achievement, that encourages risk, invention, and discovery. From our country's beginning, as I was told in history classes, what mattered in this new country was not who you were, but what you could do. Such was the ethic of the American frontier, and such is still the ethic of the scientific frontier.

The danger to our scientific soul is not that a renegade movement like intelligent design will threaten to show that the scientific establishment has got it all wrong. That sort of thing happens all the time. One can, after all, disagree with mainstream science, conclude that it is profoundly mistaken, and then set out to correct it by scientific means. Revolutions of that sort are actually endorsements of the scientific process, and an accepted part of the self-correcting (or self-affirming) nature of the discipline. If the intelligent design movement wins out over evolution by the weight of experimental evidence, logically interpreted, science will be stronger for it.

The danger, therefore, doesn't come from a scientific challenge. It doesn't emerge from threats to Darwinian ways of thinking, or rivals to evolutionary explanations for behaviors, structures, or biochemical pathways. Rather, the true danger stems from the tactics and techniques the ID movement has chosen to employ in its assault on science. As we've seen, the core strategy of the movement is to dethrone evolution by undermining any notion that science is a genuine pathway to help us grasp the true nature of existence. Its rhetoric has centered on "fairness," emphasizing that evolution is just a theory, and therefore the "theory" of design has just as much right to a place in classroom, curriculum, and textbook. Darwinism then becomes merely a story embraced by secular humanists, and intelligent design a story for people of faith. Science would no longer be a way of choosing between stories, a means of testing hypotheses, but rather a passive vehicle in which selected facts and explanations are marshaled in support of one story or the other.

At a stroke, the ID strategy transforms science. The discipline of Einstein, Galileo, Darwin, and Mendel is no longer driven by the unforgiving test of nature, but by the relative values of our times and of people who attempt to dominate and dictate cultural norms. Evolution's primary flaw is not that it is wrong, but that it can be used by groups whose cultural norms and values some believe should be rejected. Science, according to the ID critique, is controlled and manipulated by dominant groups who suppress dissent and permit only the studies and experiments that will help maintain their domination.

One cannot be certain of the long-term effect of the ID critique of science. But there is no mistaking the transforming effect the ID movement may have—in fact, is having—on the American view of science. To the partisans of the movement, changing the public perception of science is the key to achieving their goals. Remember the great "comfort" of the story of design. To the champions of ID a fundamental change in how we regard science would be a wonderful thing. It would assure us that we lead lives of meaning, value, and purpose. It would connect us with our maker by means of science, and it would recenter us in a universe in which those awful Darwinists caused us to lose our way.

But it would also reduce science to just another relativistic discipline. It would tell us that thinking the right spiritual thoughts is essential to the scientific process, and that there are no absolutes in nature. It would break the universal, democratic claim that science has on us as a people, and make it the province of the "enlightened," of the elite, for the victorious advocates of ID would become an elite in every sense. Their victory, however, would be hollow in every sense—because to gain that victory they would have debased and sacrificed our scientific soul.

Our competitors in the world don't accept the relativistic critique for a second. In fact, they have learned the lesson well that what one must bring to the table to make science work is an "American" set of values: practical, nonideological, nonsectarian, and measured only by the utility of one's ideas and theories. If we unlearn this lesson, the balance of these times will truly have tipped, and the age of American leadership in science may well have come to an end.

THE BEST OF TIMES

As more and more of my colleagues become aware of the ways in which the ID movement has sought to redefine science, they have begun to appreciate the seriousness of the threat. At first they may have underestimated antievolutionism by believing that it was restricted to a segment of society they could dismiss as uneducated and unsophisticated. But when they observed the Kansas Board of Education case, they realized that the threat was much broader. In 1999 the opponents of evolution in Kansas did the honest thing: They were against evolution, so they simply took it out of their state's standards altogether. That made Kansas a laughingstock, so in their second try (in 2005) they did something more subtle, but far more radical: They redefined science itself.

But deep within all of the doom and gloom about the scientific enterprise is a glimmer of hope—in fact, maybe something much brighter than a glimmer. Dickens's drama of the French Revolution pitted heroism against betrayal, loyalty against greed, idealism against cynicism. Its most desperate moments teetered on the abyss of loss, but had the capacity to turn suddenly into the righteousness of triumph. Even amid the very worst of times, there remains the possibility for the best in humanity. In the very depths of despair, there are the seeds of victory.

So it is today with the struggles over evolution. The tendency of Americans to reject evolution is depressing to many scientists, but it also presents a tremendous and largely unappreciated opportunity. I see a bit of this every time I speak in public, give a college lecture, or answer questions on a radio talk show. The very mention of the topic of evolution brings Americans out in huge numbers. A few years ago I gave the title "Time to Abandon Darwin?" to more than a score of talks I gave on college campuses. The audiences these talks drew were enormous, often exceeding a thousand people, who crowded into lecture rooms to hear someone question whether it was time to toss old Charles Darwin into the dustbin. I quickly assured the curious that I expected we would have Charlie to kick around for many more years, and then launched into an analysis of the nature of the evidence for evolution by natural selection. The crowds did not always like what I had to say, but they were transfixed by the subject. Americans love science, and they are fascinated even by a science that more of them reject than accept.

Therein lies the opportunity; right there is a point of light in the midst of the darkness. Americans are genuinely curious about the subject. They hunger for understanding, and they carry within them that unspoken American confidence that they, too, can pass judgment on even the most difficult scientific issues. *I don't care what the experts say,* we Americans always seem to think, *I'm going to decide for myself.* That's the mark of a nation that loves science, and we still have it in us.

That overpowering interest is a tremendous opportunity for all of us in the scientific community. It means that the ground is still fertile, and if we can step outside our laboratories and field stations for long enough, we will find a public eager to embrace the practical realities of scientific reason. But we have got to make the case, and we have got to make the effort to get that case into the public dialogue. Sadly, on this point, it is the scientific community that has let our country down. We simply have not done the job we should of taking our case to the public, of conveying the passion and strength and logic of our ideas in terms accessible to ordinary Americans. That has to change, and the change needs to start now. But the opportunity is there.

An even greater opportunity can be found, paradoxically, in the territory that design has claimed for its own—in the language of meaning and values. ID's greatest selling point is that it seems to make sense of things. It tells us, in the language of "Desiderata," that we are children of the universe—*that we are meant to be here.* There is a plan to things, and that plan includes us in the design of life.

That sounds like a comforting message, but beyond it, ID has almost nothing else to say. It cannot tell us why our bodies are "designed" the way they are, and it has no explanation for the patterns of the fossil record or our similarities to other organisms—except to claim that that was simply the way the designer chose to make us. The "science of design" asks us to marvel at the suitability of the universe to sustain life, but it scoffs at any suggestion that it might also be able to evolve that life. And design, above all, imagines life as brittle and unchanging, unable to innovate without injections of intelligence from the designer. Living organisms are not the great winners of billion-year-long contests for survival and adaptation, but just the arbitrary constructs of the moment. They lack ancestry, they lack history, and they are nothing more than the current choices of a cosmic intellect, crafted by design and then placed on our planet for purposes unknown. We cannot know the plan of this great intelligence, and design "science" cannot even identify its source.

I once heard one of my colleagues despairingly point to the explanation from design and wonder "How can genuine science compete with that?" What stories can we tell that would match their fables, that would outdo their fantasies? I think the answer is clear. We can tell the real ones.

The story evolutionary science can tell is grander and more sweeping than any just-so narrative concocted by the pretenders of intelligent design. Evolution tells us that we have a history on this planet, a history we share with every living organism. Our ancestors survived the great extinctions that nearly snuffed out life on planet Earth. They found a body plan that could produce limbs adapted for walking, running, climbing, swimming, and even flying. These adaptations explain why even today the genes that produce our forelimbs are the same ones that control the development of fins. They explain why the same DNA sequences that tell human cells to become photoreceptors will produce eyes in a fly, and why the same proteins that control cell division in yeasts will work in humans. We not only know where we came from, but increasingly we know how we got here, too.

Evolution is not just a better story, a drama with more plot twists and cliffhangers than design could ever imagine, but it has the added advantage of actually being true. It's more than a clever turn of phrase or a crafty way of looking for gaps in understanding into which we can plug a "designer." Evolution is a powerful and expanding theory that unites knowledge from every branch of the life sciences. Paleontologists now sit down with geneticists and developmental biologists to compare notes, and they find that changes over geologic time can be explained by the very genes that regulate development and growth today. Evolution draws all of biology into a single science. And that's one heck of a story.

I have no doubt that the battle over evolution will continue for years, and that it will spread from America to many other countries. In every case the question it raises will be the same: Are we willing to allow science to work? Do we have the strength and the wisdom to allow science to discard the ideas that don't work, and to search for genuine truth about the natural world? To be sure, this requires a certain degree of faith, a faith that there is an objective reality to nature, and the faith that such a reality is indeed worth knowing. There is risk in embracing faith, even faith in reality, even the faith of a scientist. But faith promises rewards as well, and in finding the strength to embrace what evolution tells us about the nature of reality, we will find reward beyond measure. For it is such faith that will ultimately redeem our scientific souls.

Notes

ONE: ONLY A THEORY

1. The case was formally known as *Selman et al. v. Cobb County Board of Education,* listed in federal court records as Civil Action No. 1:02-CV-2325-CC. It was originally heard in the Atlanta Division of United States District Court for the Northern District of Georgia. The trial described here ended in victory for the plaintiffs, and a court order was issued requiring the removal of the textbook stickers. The stickers themselves were taken out prior to the beginning of school in 2005. However, the judgment was then appealed to federal circuit court by the board of education. The appeals court did not reverse the decision, but did return the case to the district court for additional findings of fact in May 2006. Facing the possibility of a new trial, in December 2006 the school board formally ended its appeal and agreed to remove the stickers permanently and to pay the plaintiffs' legal costs.
2. Those forces were indeed successful in the elections of 2006, which returned a 6–4 proevolution majority to the Kansas Board of Education. Kansas remains one of the most contentious battlegrounds in the struggle over evolution, and at this writing a new battle over Darwin was looming for the 2008 elections.
3. Quoted in the August 3, 2005, edition of the *Washington Post,* President George W. Bush said: "Both sides ought to be properly taught... so people can understand what the debate is about." He added: "Part of education is to expose people to different schools of thought.... You're asking me whether or not people ought to be exposed to different ideas, and the answer is yes."

4. These words were taken from the narration of a publicity trailer for the 2001 *NOVA* series on evolution. There were seven programs in the series, and this narration appeared in the introduction to the final program, "What About God?"
5. My rough calculations of Nobel Prizes for the past three decades—beginning 1975, 1985, and 1995—are: Physics 67 percent, 48 percent, 71 percent; Medicine 62 percent, 75 percent, 61 percent; Chemistry 47 percent, 63 percent, 58 percent. In 2006 American scientists received *all* of the Nobel Prizes in science.
6. I would strongly recommend Wood's remarkable history of the birth of our republic *The Radicalism of the American Revolution* (New York: Vintage, 1993). His more recent books include *The American Revolution: A History* (New York: Modern Library, 2003) and *Revolutionary Characters: What Made the Founders Different* (New York: Penguin, 2006).
7. The first violent act of the Revolution was the burning of the British revenue ship HMS *Gaspee* in 1772. Although it is little known or celebrated outside Rhode Island, this act predated the Boston Tea Party (1773) by more than a year. The street circling the Rhode Island state capitol building is named Gaspee Street in memory of the event.

TWO: EDEN'S DRAFTSMEN

1. William Paley, *Natural Theology: Or, Evidences of the Existence and Attributes of the Deity,* 12th ed. (London: J. Faulder, 1809), 1.
2. Ibid., 3.
3. Ibid., 17–18.
4. Charles R. Darwin, *The Autobiography of Charles Darwin, 1809–1882,* ed. Nora Barlow (London: Collins, 1958), 59. http://darwin-online.org.uk/content/frameset?itemID=F1497&viewtype=text&pageseq=59.
5. This quotation comes from Darwin's letter of November 15, 1859, to John Lubbock, in *The Life and Letters of Charles Darwin,* ed. Francis Darwin, vol. 2 (New York: Appleton, 1887), http://charles-darwin.classic-literature.co.uk/the-life-and-letters-of-charles-darwin-volume-ii/ebook-page-08.asp.
6. Charles R. Darwin, *On the Origin of Species,* 6th ed., 1859 (New York: Oxford University Press, 1996), 187.
7. John Ray, in his book *The Wisdom of God Manifested in the Works of Creation* (1691), had earlier argued that the eye revealed the wisdom and power of God.
8. Research has only strengthened this conclusion, and two recent studies serve as examples. The first, by Gavin Young of the Australian National University, described fossils from the Devonian period that preserved some of the soft tissue around the eyes of placoderm fish. The detailed structure of this tissue provides evidence for a true intermediate form between jawless and jawed fish, documenting a critical step in the evolution of the vertebrate eye. [Gavin C. Young, "Number and Arrangement of Extra-

ocular Muscles in Primitive Gnathostomes: Evidence from Extinct Placoderm Fishes," *Biology Letters* 4 (2008): 110–14.] The second, more ambitious analysis appeared in *Nature Reviews.* Trevor Lamb (also of the Australian National University) and two of his colleagues presented a detailed explanation for the evolution of the eye, dealing with photoreceptor proteins, the structure of the retina, and the connections of the eye to the brain. These authors presented a six-stage scenario for the evolution of the organ consistent with current data, and also described a number of ways in which their scenario might be put to the test. While not all of the tests they proposed have been done, it is no longer possible to claim that evolution cannot provide a reasonable pathway to Paley's organ of extreme perfection. [Trevor D. Lamb et al., "Evolution of the Vertebrate Eye: Opsins, Photoreceptors, Retina and Eye Cup," *Nature Reviews Neuroscience* 8 (2007): 960–76.]

9. Michael J. Behe, *Darwin's Black Box* (New York: The Free Press, 1996), 214.
10. Bruce Alberts, "The Cell as a Collection of Protein Machines," *Cell* 92 (1998): 291–94.
11. Michael J. Behe, "Intelligent Design Theory as a Tool for Analyzing Biochemical Systems," in *Mere Creation: Science, Faith and Intelligent Design,* ed. William Dembski (Downers Grove, IL: InterVarsity Press, 1998).
12. Percival William Davis and Dean H. Kenyon, *Of Pandas and People: The Central Question of Biological Origins,* 2nd ed. (Dallas: Haughton, 1993), 141.
13. Behe, *Darwin's Black Box,* 87.
14. Ibid., 86.
15. David L. DeRosier, "The Turn of the Screw: The Bacterial Flagellar Motor," *Cell* 93 (1998): 17–20.
16. This quotation was reported in the *York Dispatch,* a local paper that covered the trial extensively: Christina Kauffman, "Scientist: Intelligent Design Is Science, Not Religion," *York Dispatch,* Nov. 4, 2005. It was also cited in Margaret Talbot's article "Darwin in the Dock," *New Yorker,* Dec. 5, 2005.
17. Michael J. Behe, "The Challenge of Irreducible Complexity," *Natural History,* Apr. 2002, 74.
18. Ibid., 74.
19. Both quotations were reported by Thomas W. Clark, "Who Wrote the Book of Life?," *The Humanist,* Sept. 1, 2000.
20. Strictly speaking, genes produce RNA molecules, many of which are then "translated" into chains of amino acids known as polypeptides. A protein is composed of one or more polypeptides (some proteins have several polypeptide subunits) and may contain other molecules as well. Some genes code for RNA molecules that have functions of their own, including transfer RNAs and ribosomal RNAs, and the newly discovered small interfering RNAs (siRNAs). In the face of such diversity, it has become more and more difficult to define exactly what a gene might be. For our purposes, a gene will be any DNA sequence that produces a functional product by copying that sequence into RNA.

21. Stephen C. Meyer, "The Origin of Biological Information and the Higher Taxonomic Categories," *Proceedings of the Biological Society of Washington* 117 (2004): 213–39.
22. William A. Dembski, *Intelligent Design: The Bridge Between Science and Theology,* (Downers Grove, IL: InterVarsity Press, 1999), 167–70.
23. Ibid. Emphasis is in the original.
24. Ibid.

THREE: EMBRACING DESIGN

1. Bruce J. MacFadden, "Fossil Horses—Evidence for Macroevolution," *Science* 307 (2005): 1728–30. I strongly recommend MacFadden's paper as required reading for anyone who doubts that the fossil record contains sufficient numbers of intermediate or transitional forms to document the evolutionary process.
2. It is, of course, possible to "rescue" the concept of irreducible complexity by redefining it. But that redefinition also destroys its usefulness for the design movement. One could reword the definition to state that an irreducibly complex machine is one in which the removal of a single part causes only the loss of the machine's *original* function. By this standard, the three-part mousetrap has lost its original function of catching mice, so it could still be considered irreducibly complex. However, the reason that irreducibly complex machines were vital to the design argument in the first place was that they were supposedly unevolvable because their component parts had no function at all. Once one admits that component parts can indeed have functions, the argument for unevolvability collapses. So, one can redefine irreducible complexity to save it from the mousetrap problem, but the rescue comes at the price of the argument itself.
3. "Motile Behavior of Bacteria," a feature article on the flagellum, was written for the January 2000 issue of *Physics Today* by Harvard biologist Howard Berg, and is available on the Web at http://www.aip.org/pt/jan00/berg .htm.
4. The results of two studies pointing to the solution of the problem were summarized by Claudio D. Stern et al. in "Embryology: Fluid Flow and Broken Symmetry," *Nature* 418 (2002): 29–30.
5. Michael J. Behe, "The Challenge of Irreducible Complexity," *Natural History,* Apr. 2002, 74.
6. Ibid.
7. Roderick M. McNab, "The Bacterial Flagellum: Reversible Rotary Propellor and Type III Export Apparatus," *Journal of Bacteriology,* 181 (1999): 7149–53.
8. Shin-Ichi Aizawa, "Bacterial Flagella and Type III Secretion Systems," *FEMS Microbiology Letters* 202 (2001): 157–64.
9. The original report on whales and dolphins, based on a biochemical analysis of their blood, was published in 1969 [A. Jean Robinson et al., "Whales and Dolphins Lack Factor XII," *Science* 166 (1969): 1420–22]. A more recent

report, based on direct molecular analysis of the whale and dolphin genomes, has now confirmed the lack of factor XII in these animals [Umeko Semba et al., "Whale Hageman Factor (Factor XII): Prevented Production Due to Pseudogene Conversion," *Thrombosis Research* 90 (1998): 31–37].

10. Paramvir Dehal et al., "The Draft Genome of *Ciona intestinalis:* Insights into Chordate and Vertebrate Origins," *Science* 298 (2002): 2157–67.
11. Nicholas J. White, "Antimalarial Drug Resistance," *Journal of Clinical Investigation* 113 (2004): 1084–92. The wording of Dr. White's actual statement is: "Resistance to chloroquine in *P. falciparum* has arisen spontaneously less than ten times in the past fifty years. This suggests that the per-parasite probability of developing resistance de novo is on the order of 1 in 10^{20} parasite multiplications."
12. Pooja Mittra et al., "Progressive Increase in Point Mutations Associated with Chloroquine Resistance in *Plasmodium falciparum* Isolates from India," *Journal of Infectious Diseases* 193 (2006): 1304–12.
13. This point is also made in a review by a group under Ian Hastings in *Science.* They note that as many as eight or nine different mutations are implicated in the development of chloroquine resistance in *Plasmodium,* and several more for resistance to other drugs. Ian M. Hastings et al., "A Requiem for Chloroquine," *Science* 298 (2002): 74–75.
14. Nicholas J. Matzke, "The Edge of Creationism," *Trends in Ecology and Evolution* 22 (2007): 566–67.
15. Jamie T. Bridgham et al., "Evolution of Hormone-Receptor Complexity by Molecular Exploitation," *Science* 312 (2006): 97–101.
16. Tonegawa won the Nobel Prize for Medicine in 1987. Although he now works at MIT's Whitehead Institute, because he was a Japanese national working in Switzerland at the time of his discoveries, I counted his prize as one of the non-American awards when I calculated my Nobel totals in chapter 1.
17. Michael J. Behe, *Darwin's Black Box* (New York: The Free Press, 1996), 130.
18. Ibid., 139.
19. Vladimir V. Kapitonov and Jerzy Jurka, "RAG1 Core and V(D)J Recombination Signal Sequences Were Derived from *Transib* Transposons," *Public Library of Science, Biology* 3 (2005): e181.
20. This case is formally known as *Kitzmiller v. Dover.* Its detailed legal reference is Federal Case 4:04-cv-02688-JEJ Document 342, filed Dec. 20, 2005. This quotation comes from *Kitzmiller v. Dover* decision, 78.
21. Ibid.
22. William Dembski, "Detecting Design in the Natural Sciences," *Natural History,* Apr. 2002, 76.
23. Ibid.
24. Incidentally, nature often makes use of prime numbers. As Tom Schneider of the National Institutes of Health has pointed out to me, cicadas employ prime numbers in their life cycles to avoid predators by emerging after 13 or 17 years of development. Both numbers are primes.
25. Dr. Schneider has made summaries of his results, as well as the ev pro-

gram itself, publicly available at http://www.ccrnp.ncifcrf.gov/~toms/paper/ev/.

26. Irfan D. Prijambada et al., "Emergence of Nylon Oligomer Degradation Enzymes in *Pseudomonas aeruginosa* PAO Through Experimental Evolution," *Applied and Environmental Microbiology* 61 (1995): 2020–22.
27. Technically, these small fragments were oligomers of 6-aminohexanoate, the monomer from which nylon is made, joined together by amide bonds.
28. Shelley D. Copley, "Evolution of a Metabolic Pathway for Degradation of a Toxic Xenobiotic: The Patchwork Approach," *Trends in Biochemical Sciences* 25 (2000): 261–65.
29. Glenn R. Johnson et al., "Origins of the 2,4-dinitrotoluene Pathway," *Journal of Bacteriology* 184 (2002): 4219–32.
30. John M. Logsdon Jr. and W. Ford Doolittle, "Origin of Antifreeze Protein Genes: A Cool Tale in Molecular Evolution," *Proceedings of the National Academy of Sciences* 94 (1997): 3485–87.

FOUR: DARWIN'S GENOME

1. These numbers were taken from Harris poll no. 52, July 6, 2005.
2. The law Scopes violated was generally known as the Butler Act, found in chapter 27 of laws passed by the Tennessee General Assembly in 1925. The actual wording of the act provided that "it shall be unlawful for any teacher in any of the Universities, Normals and all other public schools of the State which are supported in whole or in part by the public school funds of the State, to teach any theory that denies the story of the Divine Creation of man as taught in the Bible, and to teach instead that man has descended from a lower order of animals."
3. Charles R. Darwin, *The Descent of Man, and Selection in Relation to Sex,* 2nd ed. (New York: H. M. Caldwell, 1874), from the conclusion to chapter 2, "On the Manner of Development of Man from Some Lower Form."
4. Ibid., ch. 6.
5. Daniel E. Lieberman, "Another Face in Our Family Tree," *Nature* 410 (2001): 419–20. The image at left in figure 4.1 is redrawn from a similar diagram in the paper. The image at right in figure 4.1 is reproduced from *On the Origin of Species,* by Charles Darwin.
6. This figure was adapted from a similar chart prepared by James Foley and placed online as part of the Talk Origins Web site: http://www.talkorigins.org/faqs/homs/compare.html. The URL for Foley's chart is http://www.talkorigins.org/faqs/homs/compare.html.

   The specific fossils used for this comparison are cited by the names and identifiers by which they are commonly known in the popular literature, and not by specific scientific nomenclature. For example, ER1470 refers to a specific skull (KNM ER 1470) discovered by Bernard Ngeneo in Kenya in 1972. The scientific classification of this specimen remains in dispute among the scientific community, which is why this skull has been identified with its specimen name rather than by scientific nomenclature.

The creationist references used by Foley for Figure 4.2 are:

Sylvia Baker, *Bone of Contention: Is Evolution True?* (Darlington, UK: Evangelical Press, 1976).

Malcolm Bowden, *Ape-Men: Fact or Fallacy?* 2nd ed. (Bromley, Kent, UK: Sovereign, 1981).

Jack W. Cuozzo, *Buried Alive: The Startling Truth About Neanderthal Man* (Green Forest, AR: Master Books, 1998), 101.

Duane T. Gish, *Evolution: The Challenge of the Fossil Record* (El Cajon, CA: Creation-Life Publishers, 1985).

———, *Evolution: The Fossils Say No,* 3rd ed. (San Diego: Creation-Life Publishers, 1979).

Marvin L. Lubenow, *Bones of Contention: A Creationist Assessment of Human Fossils* (Grand Rapids, MI: Baker Books, 1992).

A. William Mehlert, "*Australopithecus* and *Homo habilis*—Pre-human Ancestors?" *Creation Ex Nihilo Technical Journal* 10, no. 2 (1996): 219–40.

David N. Menton, *The Scientific Evidence for the Origin of Man* (St. Louis: Missouri Association for Creation, 1991).

Paul S. Taylor, *The Illustrated Origins Answer Book,* 4th ed. (Mesa, AZ: Eden Productions, 1992).

———, "Who's Who and What's What in the World of 'Missing' Links?" 1996, http://www.christiananswers.net/q-eden/edn-c008.html (earlier version, 1995, with Mark Van Bebber).

7. In addition to the billions of DNA bases in our chromosomes, there is a small genome of several thousand bases carried in the DNA molecules in our mitochondria, tiny organelles within our cells that transform and release chemical energy.
8. Since DNA is double stranded, the *bases* are actually *base pairs.* However, I've tried to make this discussion understandable even to readers who aren't familiar with DNA's double-stranded complementarity, so I will refer to DNA sequences as being composed of bases rather than base pairs.
9. There's a great story, incidentally, as to why we have so many copies of these genes. As figure 4.3 shows, some of them are switched on during embryonic development, and they help to produce a hemoglobin molecule that has a higher affinity for oxygen than ordinary hemoglobin, which is exactly what a developing fetus needs in order to pull oxygen out of the blood of its mother. After birth, the adult copies of the genes are gradually switched on, making a smooth transition from life in the womb to life as an air-breathing human being.
10. The source for the analysis presented in figure 4.4 is L.-Y. Edward Chang and Jerry L. Slightom, "Isolation and Nucleotide Sequence Analysis of the

B-type Globin Pseudogene from Human, Gorilla, and Chimpanzee," *Journal of Molecular Biology* 180 (1984): 767–84.

11. Francis C. Collins, *The Language of Life* (New York: The Free Press, 2006).
12. Tarjei S. Mikkelsen et al., "Initial Sequence of the Chimpanzee Genome and Comparison with the Human Genome," *Nature* 437 (2005): 69–87.
13. LaDeana W. Hillier et al., "Generation and Annotation of the DNA Sequences of Human Chromosomes 2 and 4," *Nature* 434 (2005): 724–31.

FIVE: LIFE'S GRAND DESIGN

1. As quoted in Michelangelo D'Agostino, "In the Matter of *Berkeley v. Berkeley*," *Berkeley Science Review,* Spring 2006, 33.
2. Stephen Jay Gould, "The Verdict on Creationism," *New York Times Magazine,* July 19, 1987, 32.
3. *Kitzmiller v. Dover* decision, 32.
4. Paul A. Nelson, "Life in the Big Tent: Traditional Creationism and the Intelligent Design Community," *Christian Research Journal* 24 (2002): 4.
5. Erwin Schrödinger, *What Is Life?* (New York: Doubleday, 1944), 20.
6. The most specific and formal application of this term to biology was given by the French philosopher Henri-Louis Bergson (1859–1941).
7. Martin Rees, *Just Six Numbers* (New York: HarperCollins, 1999).
8. Walt Whitman, "Song of Myself," in *Leaves of Grass,* 1st ed. (self-published, 1855; New York: Viking, 1959) §31, l. 662.
9. So far as I am able to tell, the term "anthropic principle" was first used in 1973 by astrophysicist Brandon Carter in the context of a talk on the work of Copernicus. While the work of Copernicus is usually credited with having removed the earth (and we who dwell upon it) from a special place in the center of the universe to a rather nondescript position, Carter pointed out that in terms of the constants of nature, we humans do have a special and perhaps a privileged place. This observation was expanded by John D. Barrow and Frank J. Tipler in their book, *The Anthropic Cosmological Principle* (New York: Oxford University Press, 1986).
10. Marc W. Kirschner and John C. Gerhart, *The Plausibility of Life: Resolving Darwin's Dilemma* (New Haven: Yale University Press, 2005).
11. Ibid., 35.
12. Ibid., 226.
13. Sean B. Carroll, *Endless Forms Most Beautiful: The New Science of Evo Devo and the Making of the Animal Kingdom* (New York: Norton, 2005), 285–86.
14. Ibid., 289.
15. It is true that Haeckel's drawings inspired several generations of incorrect artwork in biology textbooks all over the world. When British developmental biologist Michael K. Richardson pointed out these errors in 1997, I was stunned to discover that even one of my own textbooks contained drawings modeled after Haeckel's. I quickly corrected them by substituting

more accurate drawings, and then by replacing those with actual photographs of the embryos.

16. Carroll, *Endless Forms Most Beautiful,* 291.

SIX: THE WORLD THAT KNEW WE WERE COMING

1. Senator Rick Santorum (R-PA), quoted from National Public Radio's August 4, 2005, *Morning Edition.* Santorum, incidentally, has been a critic of evolution throughout his career in politics. He sponsored the so-called Santorum Amendment that was briefly attached to the No Child Left Behind Act, and he was an early supporter of the Dover, Pennsylvania, school board's attempts to establish an intelligent design curriculum. He was defeated in his bid for reelection in November 2006.
2. Statement from the Answers in Genesis Web site, Oct. 2006. The current URL of this page is http://www.answersingenesis.org/get-answers/features/you-matter-to-god.
3. As quoted by Daniel Lazare in "Your Constitution Is Killing You," *Harper's,* Oct. 1999, 57–65.
4. The exceptions include, of course, conditions such as Down syndrome, technically known as trisomy-21, in which two copies of a chromosome manage to find their way into a reproductive cell. If such a cell participates in fertilization, the resulting individual will possess three copies of chromosome 21.
5. For the sake of the reader, I've left out the equally "random" story of how my own mom and dad met.
6. By "nearly every living thing" I mean only those that employ a sexual process in reproduction that involves some form of meiosis. That leaves out most prokaryotes (bacteria) and many other organisms, but it includes nearly all animals and plants, as well a large number of other organisms, such as most fungi, many algae, and many single-cell eukaryotic organisms.
7. For readers who might be, for some reason, interested in specific details, at this writing I have umpired high school and college fastpitch softball in New England for a decade and a half. Most of the college games I work are Division II and III, but I have also worked Division I games and umpired in the National Pro Fastpitch league. I also umpire youth tournaments nearly every weekend over the summer months, averaging more than one hundred games per year on the ball field.
8. Stephen Jay Gould, in the introduction to Carl Zimmer, *Evolution: The Triumph of an Idea* (New York: HarperCollins, 2001).
9. This does not mean that placental mammals were entirely absent from Australia. A number of such animals have been discovered among the early fauna of the continent. See, for example, Henk Godthelp et al., "Earliest Known Australian Tertiary Mammal Fauna," *Nature* 356 (1992): 514–16.
10. Simon Conway-Morris, *Life's Solution: Inevitable Humans in a Lonely Universe* (New York: Cambridge University Press, 2003).
11. John F. Haught, *God After Darwin: A Theology of Evolution* (New York:

Perseus Books, 1999); *Responses to 101 Questions About God and Evolution* (Mahwah, NJ: Paulist Press, 2001).

12. See, in particular, chapters 2, 3, and 4 of my *Finding Darwin's God: A Scientist's Search for Common Ground Between God and Evolution* (New York: HarperPerennial, 2007).
13. From *Discourses of the Popes from Pius XI to John Paul II to the Pontifical Academy of Sciences, 1936–1986* (Vatican City: Pontificia Academia Scientiarum, 1986), 161–64.
14. A more complete quotation is: "Even a non-Christian knows something about the earth, the heavens,...the kinds of animals, shrubs, stones, and so forth, and this knowledge he holds to as being certain from reason and experience. Now it is a disgraceful and dangerous thing for an infidel to hear a Christian, presumably giving the meaning of Holy Scripture, talking nonsense on these topics; and we should take all means to prevent such an embarrassing situation, in which people show up vast ignorance in a Christian and laugh it to scorn." St. Augustine, *On the Literal Meaning of Genesis* 1.19, trans. Johannes Quasten, Walter S. Burghardt, and Thomas Comerford Lawler, *The Literal Meaning of Genesis,* vol. 1, ed. John Hammond Taylor, Ancient Christian Writers series (Mahwah, NJ: Paulist Press, 1982).
15. St. Augustine, as quoted on page 115 of *Just Six Numbers,* by Martin Rees. A similar quotation from Augustine appeared in Howard J. Van Till, "Basil, Augustine, and the Doctrine of Creation's Functional Integrity," *Science and Christian Belief* 8, no. 1 (Apr. 1996): 21–38.
16. Benedict XVI, Homily from Mass for the Inauguration of the Pontificate of Pope Benedict XVI, Apr. 24, 2005.
17. As quoted in John D. Barrow and Frank J. Tipler, *The Anthropic Cosmological Principle* (New York: Oxford University Press, 1986), 318.
18. This quotation from St. Augustine comes from his *Reply to Faustus the Manichean,* 26.3. I am very grateful to Dr. James Miller for pointing out this quotation to me.
19. There are many recent examples, of course, including Richard Dawkins, *The God Delusion* (Boston: Houghton Mifflin, 2008); Daniel C. Dennett, *Breaking the Spell: Religion as a Natural Phenomenon* (New York: Penguin Books, 2007); and Sam Harris, *Letter to a Christian Nation* (New York: Knopf, 2006).
20. Epistle of James 1:18.

SEVEN: CLOSING THE AMERICAN SCIENTIFIC MIND

1. Allan M. Bloom, *The Closing of the American Mind* (New York: Simon & Schuster, 1987), 25–26.
2. Ibid., 38.
3. Ibid.
4. Ibid., 260.
5. Ibid., 356.

6. Ibid., 358n.
7. Ibid., 373–74.
8. These admissions came in the form of documents posted on the institute's Web site. In 2006 the institute posted a page entitled "The 'Wedge Document': So What?" This posting acknowledged authorship of the document but downplayed its significance. The Wedge document is available at this book's Web site, http://www.onlyatheorythebook.com.
9. From the Wedge document.
10. Ibid.
11. Phillip Johnson, quoted in the *Berkeley Science Review,* Spring 2006, 33.
12. Paul Nelson, quoted in "The Measure of Design," *Touchstone* 17, no. 6 (2004): 64–65.
13. *Kitzmiller v. Dover* decision, p. 88.
14. Claudia Wallis, "The Evolution Wars," *Time,* Aug. 15, 2005.
15. The full title of the law was Balanced Treatment for Creation-Science and Evolution-Science Act. It was declared unconstitutional in 1987 by the United States Supreme Court in the case of *Edwards v. Aguillard.*
16. Karl Popper, *The Open Society and Its Enemies,* vol. 2 (London: Routledge, 1945), 369ff.
17. Paul K. Feyerabend, *Against Method: Outline of an Anarchistic Theory of Knowledge* (London: Verso Press, 1975), 295.
18. From the Wedge document.
19. Kansas Science Standards, Minority Report, p. 15. The Minority Report was never formally published. However, a copy of it was placed online at a Web site maintained by pro-ID forces: http://www.kansasscience2005.com/. A copy of it is also available at this book's Web site. http://www.onlyatheorythebook.com.
20. Kansas Science Standards, Minority Report, p. 25.
21. Ibid., 2.
22. Ibid., 5.
23. Ibid., 4. The Minority Report quotes from the standards existing at the time.
24. Ibid.
25. Ibid., 5.
26. Actually, there are twenty-three occurrences of the noun "naturalism" or the adjective "naturalistic" in the report.
27. Minority Report, 6.
28. William A. Dembski, *Intelligent Design: The Bridge Between Science and Theology* (Downers Grove, IL: InterVarsity Press, 1999), 14–15.

EIGHT: DEVIL IN THE DETAILS

1. Actually, in recent years the intelligent design movement has made substantial inroads in Europe, including in the United Kingdom. Intelligent design movements are active throughout Europe and have also assumed

some prominence in the Islamic world. These developments were reported in "In the Beginning," *The Economist,* Apr. 19, 2007.

2. The origins of the movement have been painstakingly documented by Barbara Forrest and Paul R. Gross in *Creationism's Trojan Horse: The Wedge of Intelligent Design* (New York: Oxford University Press, 2004). If one cares to look for the marks of conspiracy, including the systematic denial of the movement's actual goals, they are certainly there.
3. The actual statement by Minnich came in his agreement to a question by attorney Steven Harvey during an exchange on the twenty-first day of the trial, November 4, 2005. Minnich's answer contained a qualification that the supernatural was indeed a possible scientific explanation for design, but so were other, unnamed, sources of intelligent action. He never suggested any sources of design other than the supernatural, so it seems appropriate to me to infer that this was indeed the meaning of his answer. Here is the exchange from the *Kitzmiller v. Dover* trial transcript:

   > [Attorney Steve Harvey] Q. For intelligent design to be considered science, the definition of science or the rules of science have to be broadened so that supernatural causes can be considered, correct?
   >
   > [Dr. Scott Minnich] A. Correct, if intelligent causes can be considered. I won't necessarily—you know, you're extrapolating to the supernatural. And that is one possibility.

4. Many ID theorists, including Phillip Johnson, have described their view of a creator or designer as "whimsical": "It seems to me that the peacock and peahen are just the kind of creatures a whimsical Creator might favor, but that an 'uncaring mechanical process' like natural selection would never permit to develop." Phillip E. Johnson, *Darwin on Trial* (Downers Grove, IL: InterVarsity Press, 1995), 30–31.

   William Dembski clearly endorses this view of the designer. On August 8, 2006, Dembski gave an example of this frame of mind on his blog, Uncommon Descent. Under a striking photograph of a high-altitude rainbow, Dembski wrote: "This is a fire rainbow—one of the rarest naturally occurring atmospheric phenomena. The picture was captured this week on the Idaho/Washington border. The event lasted about one hour. Clouds have to be cirrus, at least four miles in the air, with just the right amount of ice crystals; and the sun has to hit the clouds at 58 degrees. It's the gratuitousness of such beauty that leads me to rebel against materialism." The URL of this posting is http://www.uncommondescent.com/intelligent-design/fire-rainbow/.
5. Forrest and Gross, *Creationism's Trojan Horse,* 16.
6. The connection of antievolution sentiments with conservative political views has been well documented by public-opinion polls. One of the most interesting of these was reported by Alan Mazur in a 2007 letter to *Science* ("Disbelievers in Evolution," *Science* 315: 187). Data presented with the let-

ter showed that self-identified political conservatives rejected evolution at much higher rates than liberals or moderates of similar educational levels and religious preference.

7. Adam Smith, *An Inquiry Into the Nature and Causes of the Wealth of Nations* (1776), vol. 1 (Indianapolis: Liberty Fund, 1981), 456.
8. George F. Will, "Grand Old Spenders," *Washington Post,* Nov. 17, 2005.
9. Charles Krauthammer, "Phony Theory, False Conflict," *Washington Post,* Nov. 18, 2005.
10. Laurie Lebo, "Dover Figures Deny Remarks on Creationism: Their Depositions Contradict What Others Remember," *York Daily Record/Sunday News,* Jan. 16, 2005.
11. Percival William Davis and Dean H. Kenyon, *Of Pandas and People: The Central Question of Biological Origins,* 1st ed. (Richardson, TX: Foundation for Thought and Ethics, 1989).
12. Books on the trial include Edward Humes, *Monkey Girl* (New York: Ecco, 2007); Matthew Chapman, *40 Days and 40 Nights: Darwin, Intelligent Design, God, Oxycontin, and Other Oddities on Trial in Pennsylvania* (New York: Collins, 2007); and Gordy Slack, *The Battle over the Meaning of Everything: Evolution, Intelligent Design, and a School Board in Dover, PA* (San Francisco: Jossey-Bass, 2007). Excellent articles on the trial were published in *Harper's* ("God or Gorilla: A Darwin Descendant at the Dover Monkey Trial," by Matthew Chapman, Feb. 2006) and in the *New Yorker* ("Darwin in the Dock," by Margaret Talbot, Dec. 5, 2005).
13. The strategy was described by William Dembski in a 2005 Web posting: http://www.uncommondescent.com/darwinism/the-vise-strategy-squeezing-the-truth-out-of-darwinists/.
14. *Kitzmiller v. Dover* decision, p. 79.
15. For the sake of those unfortunate enough to have been born recently, Mr. Wizard was a character assumed by an actor and science enthusiast named Don Herbert. Mr. Herbert hosted a weekly television show on science and technology aimed directly at American children. He invited young people from his television "neighborhood" into his house (something sadly unthinkable today) and demonstrated basic principles of science and engineering in an entertaining way. I know very few scientists of my own generation who did not watch *Mr. Wizard* as kids. For more than twenty years, Mr. Herbert introduced American children to the wonders and delights of scientific inquiry. He passed away on June 12, 2007.
16. Jon D. Miller, Eugenie C. Scott, and Shinji Okamoto, "Public Acceptance of Evolution," *Science* 313 (2006): 765–66.
17. Harris poll, June 17–21, 2005. N = 1,000 adults nationwide; margin of error: ± 3 percent.

Index

Page numbers in *italics* refer to figures.
Page numbers beginning with 223 refer to endnotes.